God, Time, Infinity

Philosophische Analyse/ Philosophical Analysis

Herausgegeben von/Edited by
Rafael Hüntelmann, Christian Kanzian, Uwe Meixner,
Richard Schantz, Erwin Tegtmeier

Band/Volume 75

God, Time, Infinity

Edited by
Mirosław Szatkowski

DE GRUYTER

ISBN 978-3-11-059193-4
e-ISBN (PDF) 978-3-11-059416-4
e-ISBN (EPUB) 978-3-11-059203-0
ISSN 2198-2066

Library of Congress Control Number: 2018941001.

Bibliographic information published by the Deutsche Nationalbibliothek
The Deutsche Nationalbibliothek lists this publication in the Deutsche Nationalbibliografie; detailed bibliographic data are available on the Internet at http://dnb.dnb.de.

Printing: CPI books GmbH, Leck
♾ Printed on acid-free paper
Printed in Germany

www.degruyter.com

FSC www.fsc.org MIX Papier aus verantwortungsvollen Quellen FSC® C083411

Contents

Acknowledgements

The publication of this book was preceded by the 3rd *International God-Oriented Conference*, with the title: *God, Time, Infinity*, which took place in Warsaw in September 2015. That conference was organized under the auspices of the Institute of Philosophy and Sociology of the Polish Academy of Sciences (IFiS PAN). I would like to thank the National Science Center for funding it, and also for the financial support of the publication of this book. Moreover, I am grateful to IFiS for acting as the unit implementing the project. Special thanks go to Ms. Grażyna Drążyk from IFiS PAN, who helped me in different ways throughout the preparation of the grant proposal to the National Science Center and the organization of the conference.

I would like to express my sincere gratitude to all thirteen authors who accepted the invitation to contribute to the volume and showed great understanding and patience during the publication process. I would also like to thank all speakers at the conference, and Dr Bartłomiej Skowron for his faithful support of my initiatives.

Last, but certainly not least, I would like to thank my brother, Sławomir Szatkowski, for his help in preparing this volume for print. I have tried to make the best possible use of all his suggestions and corrections.

Mirosław Szatkowski

https://doi.org/9783110594164-201

Andrea C. Bottani and Riccardo Fedriga

Ockham, Plantinga and the Row of Ants

1

For millennia, philosophers have discussed whether divine omniscience is compatible with human freedom – conceived of in a libertarian way – or not. If libertarianism is true, some actions are free and no action is free unless it is within the agent's power to act otherwise. If God is omniscient, however, He completely knows how I will act in the future, which seems to entail that it is never within my power to act otherwise, provided I cannot change God's past beliefs. Therefore, I am not free in the libertarian sense, or so it seems. In a nutshell, the problem is that every action that follows from a free choice entails a contingent view of the future, which nonetheless appears fixed to the constraint of the divine knowledge that foreknows it.[1]

This problem has a long history. It was addressed, discussed and purportedly solved by a number of authors, among which Augustine, Boethius, Anselm, Aquinas, Ockham, Molina and various open theists, and has raised renewed interest after Nelson Pike's article "Divine Omniscience and Voluntary Action" (Pike (1965)).[2] In his *Tractatus de Praedestinatione et de Praescientia Dei respectu futurorum contingentium* (see, Ockham (1969)), Ockham famously defended the compatibility of divine omniscience and human freedom by distinguishing between past propositions *secundum vocem* and *secundum rem*. In the words of Ockham himself:

1 With the term "fatalism" we refer to the idea that it is not in our power to act differently from the way we act; if we combine this idea with the notion of an intelligent, divine being which is defined by certain essential attributes (perfection, omnipotence) and, therefore, with the existence of certain propositions of a particular epistemic nature (infallibility, omniscience) that bind future states, we obtain the specific form of fatalism that goes under the label of "theological fatalism". See, Normore (1985) and Zagzebski (2011). See also, Kane (2002); Fischer *et al.* (2009) and Rice (2010).

2 It is not possible to reconstruct the whole debate on compatibilism and the openness of God here: the bibliography is vast and the discussion is by no means over. For a summary of the various positions, cf., Fischer (1983); Craig (1986); Craig (1991); Zagzebski (1991); Fischer *et al.* (2007); Fischer (2011), pp. 165–95; Fischer *et al.* (2007). For a historical/theoretical reconstruction, see, Craig (1988); Marenbon (2005) and Marenbon (2013); see also, for the debate from Duns Scotus to Leibniz, Michon (2004); Mugnai (2013); Fedriga (2015). For an overview of Open Theism, see, Pinnock *et al.* (1994).

https://doi.org/9783110594164-001

> Some propositions are about the present as regards both their wording and their subject matter. Where such [propositions] are concerned, it is universally true that every true position about the present has [corresponding to it] a necessary one about the past – e.g., 'Socrates is seated', 'Socrates is walking', 'Socrates is just' and the like. Other propositions are about the present as regards their wording only and are equivalently about the future, since their truth depends on the truth of propositions about the future. Where such [propositions] are concerned, the rule that every true proposition about the present has [corresponding to it] a necessary one about the past is not true.[3]

While propositions that truly refer to the past (*secundum rem*, as " Socrates was sitting" are fixed and unchangeable, others that refer to the past only verbally (*secundum vocem*, as " Peter was predestined") remain epistemically undetermined, because they are waiting to receive their proper truth-value from a state of affairs happening in the future. All foreknowledge acts fall under the latter category. Far from fixing the future, they receive their truth-value from future facts. As Nelson Pike puts it, they describe "soft" rather than "hard" facts.[4] Soft facts about the past are not properly about the past and so they fail to be "accidentally necessary". Since William of Sherwood (XIII century), medieval philosophers and theologians call *necessitas per accidens* the principle according to which, if a certain event or state of affairs is in the past, then there is nothing that one could do about it *now*.[5] In common words: what's done it is done; what has been has been. Ockham's idea is that past divine foreknowledge, however past, is not accidentally necessary, because it is about the future. So it is within our power to act in such a way that God would not have believed what in fact he does. In Fischer's words, Ockham's intuition is this:

> Do not think of the past fact that Jones would write the paper as *forcing* Jones to write the paper, or *constraining* what Jones has it within his power to do. Rather, think of Jones' free

---

3 Guillelmus de Ockham (1978a), q. I, III Propositio, ll, 208–216, p. 515 (english translation quoted in: Ockham (1969), pp. 46–47):

> Aliquae sunt propositiones de praesenti secundum vocem et secundum rem, et in talibus est universaliter verum quod omnis propositio de praesenti vera habet aliquam de praeterito necessariam, sicut tales: "Sortes sedet", "Sortes ambulat", "Sortes est iustus", et huiusmodi. Aliquae sunt propositiones de praesenti tantum secundum vocem et sunt aequivalenter de futuro, quia earum veritas dependet ex veritate propositionum de futuro; et in talibus non est ista regula vera quod omnis propositio vera de praesenti habet aliquam de praeterito necessariam.

4 Cf., Sanders (1966); Pike (1966); Adams (1967).

5 See, Knuuttila (2013); Freddoso (1983), Maierù (1972). For the medieval notion of necessitas per accidens it occurs for the first time in a logical context in Guillelmus de Sherwood (1983), (1.7.1).

decision to write the paper as *explaining why it was true* that Jones would write the paper. In other words, says the Ockhamist, Jones' free decision to write the paper is the *explanatory ground* of the fact that, at t_1, it was true that he would write the paper.[6]

Ockham's solution faces a number of problems. Among others, both the notion of accidental necessity and the related concept of a hard fact stand in need of a more precise definition; it is far from clear what it is for an action to be within one's power; and the purported "way out" seems to presuppose eternalism, being ultimately unavailable to those who are sympathetic to some anti-eternalist conception of time (since the "softness" of divine foreknowledge precisely consists in its dependence on the future foreknown fact, which requires the reality of the future).

In his famous paper "Ockham's Way Out" Plantinga strived to clarify accidental necessity by completely detaching this notion from the idea that the past is fixed.[7] It is not by contrasting the closeness of the past with the openness of the future that one can get clearer about *accidental necessity*, because the future is just as unchangeable as the past. Consider that, in order to change the past, one should be able to do an action A at t_1 such that a proposition P, referring to the past and true at t_1, becomes false at t_2, after A is done, which is impossible. But then, changing the future must be equally impossible, for nobody is able to do at time t_1 an action A such that before doing A, a proposition P regarding time t_2 was true and after doing A that proposition is false. Take for example the proposition "Jones will write his article tomorrow at 7 p.m.". If the proposition is true at 3 p.m., nobody can do anything at 4 p.m. that makes the proposition false after 4 p.m. Since the future is no less unchangeable than the past, accidental necessity is not to be identified with unchangeability and has nothing to do with the asymmetry in rigidity that past and future display.[8]

Having discarded temporal asymmetry as a criterion to define accidental necessity, Plantinga manages to define the notion in terms of *power of an agent*. If Jones was born in 1967, there is nothing in my power that I could do today to prevent that event from happening or to modify it in any way. So, that event is neces-

6 Fischer *et al.* (2009), p. 256.

7 Plantinga (1986).

8 For the asymmetry of the past as the ground for transferring the strength of the necessity per accidens to the present and the future and for a solution that combines Aquinas with Ockham, Zagzebski (2011). For a historical and theoretical critique to this position, which too readily smoothes over the differences between Ockham's Contingentist point of view and Aquinas "theological determinism", see, Fedriga (2015), pp. 126–131.

sary *per accidens*.[9] It is outside of my power, or so it seems. Here is a first tentative definition along these lines:

> D) p is accidentally necessary at t if and only if p is true at t and it is not possible both that p is true at t and that there exists an agent S and an action A such that (1) S has the power at t or later to perform A, and (2) if S were to perform A at t or later, then p would have been false.

The important point is that, in the defined sense, past acts of divine foreknowledge are not accidentally necessary. Indeed, as Plantinga argues: 1) Backtracking counterfactuals of the form "If X had refrained from doing Y at t_2 then God would not have believed at t_1 that X would do Y at t_2" ($t_1 < t_2$) are true. 2) For many actions Y and many agents X, it is within X's power to refrain from doing Y.

According to Plantinga, it is the counterfactual power of action that works as a signal of a missing accidental necessity.[10] There are complications, however, and Plantinga imagines a case where a future action would be such that, were it to happen, even genuine past propositions (*secundum rem*, in Ockham's words) referring to a *hard fact* would be false. Let us suppose that a colony of carpenter ants moved into Paul's yard last Saturday; if the ants were to remain and Paul were to mow his lawn this afternoon, the colony would be destroyed. However, God intends that it be preserved. Paul will not mow his lawn this afternoon and God, who is omniscient, knew in advance this fact; but if he had foreknown instead that Paul would mow this afternoon, then he would have prevented the ants from moving in. So if Paul were to mow his lawn this afternoon, then the ants would not have moved in last Saturday. But it is within Paul's power to mow this afternoon: if Paul has this power, then there is an action (mowing the lawn this afternoon) such that if he were to perform it, the proposition

> (E) That colony of carpenter ants moved into Paul's yard last Saturday

would have been false. But (E) appears to describe a true *hard fact*, because it is strictly about the past. In the defined sense, (E) would not have *necessitas per*

---

9 Plantinga (1986), p. 247. For Plantinga's understanding of necessity, see, Plantinga (1974).
10 Cf., Plantinga (1986), p. 253:

> p is accidentally necessary at t if and only if p is true at t and it is not possible both that p is true at t and that there exists an agent S and an action A such that (1) S has the power at t or later to perform A, and (2) if S were to perform A at t or later, then p would have been false.

accidens: Plantinga concludes that not all true propositions strictly about the past (not even *hard facts*) are accidentally necessary in the defined sense.

Unfortunately, this kind of example can be reiterated in case of virtually all past objects and events: every individual *S* has the abstract power to perform an action *A* such that, if *S* were to do it, God, having foreseen it, would have refrained from creating a certain being *Z*, or from letting a certain event *F* happen. As a consequence, propositions *strictly about the past* (with a relation of strict implication) such as "*Z* existed" and "*F* happened" fail to be accidentally necessary and the facts that fall under accidental necessity in the defined sense become so rare that the definition seems to be hardly acceptable.[11]

In order to solve the difficulty, Plantinga drastically restricts the actions that an agent can perform in order to prevent non-accidentally necessary past objects or events from taking place to those he calls "basic actions" – where *A* is a basic action just in case an agent *S* can perform *A directly*, i.e., without having to perform another action *B* as a means to perform *A*.[12] The issue is complex and controversial, so let us comment but cursorily on the difficulties faced by this proposal. A first problem is posed by the elusiveness of the notion of "basic action". Another problem is that Plantinga is not completely clear about what he understands with our power to act so as to prevent a soft fact about the past (in particular, a divine act of foreknowledge) from taking place. As Hasker argues, to say that there are no compelling external forces preventing the agent from acting in that way is a *petitio principii* (it amounts to assuming that the past divine act of foreknowledge is not such a force). According to Plantinga, *S* has the power to act freely at t_2 even if at t_1 God already knows how *S* will act at t_2, because *S* has the power to act at t_2 in such a way that, if *S* were to act in that way, God would not have believed at t_1 what He does (which means that His past foreknowledge is not accidentally necessary). However, Hasker argues, if the problem is showing that divine foreknowledge does not prevent our free will (understood in the libertarian way as the capacity of acting without any constraints upon one's will), then free will is the *demonstrandum* and cannot be used as a premise, which is exactly what Plantinga seems to do. Indeed, Plantinga's argument seems to go as follows: action *A* of *S* at t_2 is free because *S* has the power to modify God's belief at t_1 in

11 Plantinga (1986), p. 254.
12 Plantinga (1986), p. 260:

> An action *A* is a basic action for a person *S* if and only if there is an action *A** that meets two conditions: first, *S* can directly perform *A**, and secondly, *S*'s being in normal conditions and his directly performing *A** is causally sufficient for his performing *A*.

such a way that action *A* is free. This would be a vicious circle, with a premise in place of the conclusion and with the *explanandum* used as the proof of itself.[13] Is there any way to escape Hasker's criticism?

2

Our aim is to defend the core of Ockham's (and Plantinga's) way-out by reinterpreting their compatibilism in new terms. We shall begin by arguing that God's omniscience is incompatible with the openness of the future. Then we will show that, despite the closeness of the future, there is a fairly coherent modal sense in which it is within our power to act differently. For this modal sense to receive full articulation, a definition of hard fact is required. This is done in non-modal terms, i.e. in terms of actual grounding. So, let us begin.

Suppose that:

- *S* is a subject;
- *A* is an action;
- "does" is an abbreviation of "does, did or will do" (the same, *mutatis mutandis*, for "is true", "knows" and the like.
- At t_2 *S* does *A*.

Assume that:

(i) the future is open: the proposition that *S* does *A* at *t* can be definitely true (or false) at t_2 and neither true nor false at t_1 ($t_1 < t_2$)

(ii) *A* proposition is a function from possible worlds *and times* to truth-values (and *not* a function from just possible worlds to truth-values). [(ii) follows from (i). If the future is open, at least some propositions must have definite truth-values at t_2 that they fail to have at t_1].

(iii) If *S* believes a proposition *P* at a time *t*, *S*'s belief is true just in case *P* is true at *t* (if I believe now that it is raining, for example, my belief is true just in case the proposition that it is raining is true now. And, if I believe now that it will rain tomorrow, my belief is true just in case the proposition that it will rain tomorrow is true now). Therefore, if at *t* *S* believes that *p*, then at *t* *S* knows that *p* just in case at *t* it is true that *p* (and, of

13 Hasker (2001), p. 103:

> How can Cuthbert have the power to cause *Cuthbert will purchase an iguana* at t_3 to be false, when its truth is immutably fixed and guaranteed by the truth of *God believes at* t_1 *that Cuthbert will purchase an iguana at* t_3?

course *S*, is justified in believing that *p*). And, if God is omniscient, at *t* God knows that *p* just in case at *t* it is true that *p* – just in case at *t* *p*.

(iv) God exists either in time or out of time (inclusive).

(v) God is omniscient.

In case God exists in time, His omniscience can consist in one of two distinct abilities:

(v_1) at every time at which God exists, God knows everything that is true at that time.

(v_2) At every time at which God exists, He knows, for every time, everything that is true at that time.

Let us say that God is omniscient$_1$ iff (v_1) is the case, while He is omniscient$_2$ in case (v_2) is the case. Obviously, omniscience$_2$ entails omniscience$_1$. If determinism were true, moreover, omniscience$_1$ would entail omniscience$_2$, for the total set of present truths would entail the total set of future truths (as well as of past truths, if one accepts some version of a strong, bi-directional determinism). In a non-deterministic world, however, omniscience$_1$ is not equivalent to omniscience$_2$ but weaker than the latter. A world where the future is open is *a fortiori* non deterministic. Therefore, in no such world omniscience$_1$ entails omniscience$_2$.

On the other hand, if the omniscient God is out of time, then, for any time, He knows everything that is true at that time, but there is no time at which He knows that, so He is neither omniscient$_1$ nor omniscient$_2$. Let us say that in such a case He is omniscient$_3$. Omniscience$_3$ is the exact atemporal analogue of omniscience$_2$.

With this battery of notions and assumptions in hand, we can prove what follows.

1) If the future is open, God is in time, and God is omniscient$_1$ but not omniscient$_2$, then at t_1 God does not know that *S* does *A* at t_2:

1) At t_1 it is neither true nor false that at t_2 *S* does *A*	From the openness of the future
2) At t_1 it is not true that at t_2 *S* does *A*	From 1
3) At t_1 God does not believe that at t_2 *S* does *A*	From 2, modulo (iii) and (iv)[1]
4) At t_1 God does not know that at t_2 *S* does *A*	From 3, by definition of knowledge

If at t_1 God does not know that at t_2 *S* does *A*, nothing in God's knowledge prevents *S* from abstaining from doing *A* at t_2. So, if God is in time and God is omniscient$_1$ (but not omniscient$_2$), His omniscience may seem to stop threatening human freedom. But it is not clear that it does. Consider that, either at t_1 it is permanently neither true nor false that at t_2 *S* does *A*, or at t_1 this "becomes" true when, with the flowing of time, t_2 becomes present. In the former case, nobody at t_1 could make a true prediction about *S*'s doing *A* at t_2. Both the proposition that at t_2 *S* does *A* and its negation are indeed permanently non-true at t_1. If *X* believes at t_1 that *S* will do *A* at t_2, and then, when t_2 becomes present, *S* does *A* at t_2, it would be incorrect to say that *X*'s belief was right, i.e., that *X*'s prediction was veridical. But this is utterly implausible. In the latter case, *X*'s belief is neither true nor false "before", and "then" true at the very same time, i.e. at t_1. This kind of paradox is implied in the very idea of changing the past, and seems to entail a contradiction. The contradiction cannot be dispelled by postulating that *X*'s belief can be both true and non-true at t_1 relative to different times (neither true nor false at t_1 relative to t_0 and true at t_1 relative to t_3, for example), for this would close the future once again. The reason is that *X*'s belief is true at t_1 relative to t_2 just in case *S* does *A* at t_2. So, if *X*'s belief is true at t_1 relative to t_2, *S* cannot refrain from doing *A*.

2) If the future is open, God is in time, and God is omniscient$_1$ but not omniscient$_2$, then what God knows at t_1 is incompatible with what He knows at t_2:

1) At t_1 it is neither true nor false that at t *S* does *A*	From the openness of the future
2) At t_1 God knows that it is neither true nor false that at t_2 *S* does *A*	From 1 and (v)$_1$
3) At t_1 God knows that it is not true that at t_2 *S* does *A*	From 2
4) At t_2 is true that *S* does *A*	By hypothesis
5) At t_2 God knows that it is true that at t_2 *S* does *A*	From 4 and (v)$_1$
6) At different times God knows that it is true and it is not true that at t_2 *S* does *A*	Fom 3 and 5

3) If the future is open, God is in time, and God is omniscient$_2$, then God's knowledge at a single time is incoherent. But no knowledge can be incoherent, by definition of "knowledge":

1) At t_2 it is true that at t_2 *S* does *A*	By hypothesis
2) God is omniscient$_2$	By hypothesis
3) At t_1 God knows that at t_2 it is true that at t_2 *S* does *A*	From 1, 2
4) At t_1 God knows that it is true that at t_2 it is true that at t_2 *S* does *A*	From 3 and '*p* iff it is true that *p*'
5) At t_1 it is neither true nor false that at t_2 *S* does *A*	From the openness of the future
6) At t_1 God knows that it is neither true nor false that at t_2 *S* does *A*	From 5, 2

7) At t_2 S does A iff at t_2 it is true that at t_2 S does A	Assumption
8) At t_1 God knows that it is neither true nor false that at t_2 it is true that at t_2 S does A	From 6, 7
9) At t_1 God knows that it is not true that at t_2 it is true that at t_2 S does A	From 8
10) At t_1 God knows that it is true and it is not true that at t_2 it is true that at t_2 S does A	From 4, 9
11) God's knowledge at t_1 is incoherent	From 10

4) If God exists outside of time, and God is omniscient$_3$, then the future cannot be open (*pace Boethius*).

Why should one endorse 4)? The so-called Boethian solution is grounded in the idea of a tenseless God roughly as follows.[14] Suppose God exists, God is outside of time and God is omniscient. Given that God is outside of time, He has no temporal states, so it is not the case that at t_1 God knows that at t_2 S will do some action A. Boethius argues that, if at t_1 God does not know that at t_2 S will do A, then at t_1 S has still the power to act differently at t_2. At t_1, S can still abstain from doing A at t_2. But it is not clear that this is the case. Consider that, even if it is not the case that at t_1 God knows that at t_2 S will do A, certainly it is the case that at t_1 it is true that God (atemporally) knows that at t_2 S does A. Similarly, it is not in time that $2 + 2 = 4$, for this is an atemporal truth. But since this is atemporally true, it is *a fortiori* true at the present time too, and at any other moment. If at t_1, and at any other moment, it is (atemporally) true that at t_2 S will do A, how could S abstain from doing A in any way? If atemporally God knows that p, then atemporally it is true that p. But if atemporally it is true that p, then at any time it is equally true that p. Consider Fine's distinction between eternal and sempiternal truths, where an eternal truth is a proposition that is true "regardless of the time" while a sempiternal truth is a proposition that is true "whatever the time". And consider that, as Fine rightly insists, all the eternal truths (as $2 + 2 = 4$) are

14 By "the so-called Boethian traditional solution" (or Boethius-Aquinas solution) we mean the thesis that all things, past, present and future, are metaphysically present to God, i.e., that God's present is co-extensive with worldly past, present and future. In addition to this reading (see, Craig (1988) and Craig (1991)), it is worth mentioning another recent and persuasive interpretation, which we may dub "the philosophical view" (Marenbon (2013)) and is based on a close reading of the Boethian texts in their context (Cons. V, 6 ff). According to the philosophical view, the lot "all things, past, present and future are present to God", should be understood epistemically, i.e. He knows them as if they were in His present simplicity. In the present context, we might understand this in terms of simple divine *presentism* in knowing things. See, Helm (2010).

a fortiori sempiternal.[15] Whether the Divine knowledge that closes the future preventing *S* from doing otherwise is located in time or outside of it, it does not seem to make a big difference for *S*'s incapacity to do otherwise.

3

The moral to be drawn from 1–4 is probably that Divine Omniscience, however coinceived of, does not seem to get smoothly along with the openness of the future. So, suppose the future is closed. What would it be of *S*'s power to do otherwise in such a case? Is there any sense in which one might be said to be able to act otherwise in case the future is closed? Well, the future might be closed even in a non-deterministic universe. A block universe might be such that none of its temporal slice (completely) determine its subsequent slices (and each is compatible, looking in the backward direction, with more than one series of preceding slices).[16] Possibility of doing otherwise might simply be logical, and physical, and chemical, and biological, and so on, compatibility of *many alternative actions*, with the same past.

This past, however, must be purified from "soft facts", including facts of Divine foreknowledge. This is necessary, for no two alternative actions can be both compatible with a past in which God knows that only one of them takes place. If God <u>knows</u> at t_1 that I <u>do</u> *A* at t_2, it <u>is</u> true at t_1 that I <u>do</u> *A* at t_2, which is (logically) incompatible with my doing *B* instead of *A* at t_2. How to give a clear-cut distinction between hard and soft facts about the past?

Suppose Riccardo builds a time machine, gets on the machine, writes "500 years back" on a quadrant and press the button 'start'. After about ten minutes the notice "here we are" appears on the quadrant just beside the phrase "500 years back". Getting off the machine, the time traveler finds himself in Duomo Square, Milan, at noon of 1st November 1614, exactly 500 years before his departure. Call '*A*' the traveler's action of pressing 'start' and '*F*' the machine appearance in Duomo Square at noon of 1st November 1614. If Riccardo had abstained from doing *A*, *F* would not have occurred. But *F* is a hard fact about the past, and it is not in Riccardo's power to bring about that a hard fact about the past had not occurred. Therefore, it is not in Riccardo's power to abstain from doing *A*.

15 Fine (2005); for the notion of Eternity, see, Helm (2010) and Kretzmann and Stump (1981).
16 By "temporal slice of the universe" we roughly mean the total state of the universe at a single time. There are of course complications with special relativity, but let us leave this aside here.

It is assumed that *F* is a hard fact about the past when Riccardo does *A*. In a sense, however, the arrival to Duomo Square at noon of 1st November 1614 is in Riccardo's future when Riccardo does *A* (not in the subjective future of his personal time, but in the objective future of the physical time in which Riccardo is located: the physical time flowing inside the machine). It is not in question, of course, that *F* occurs 500 years before *A*: it does, but only outside the machine. Inside the machine, however, the temporal order is quite reversed, and *F* occurs ten minutes after *A*. It looks as if there were two temporal orderings. Many events, for example the Waterloo battle, have a place in the former ordering but not in the latter. Others, for example Riccardo yawning during the time journey, have a place in the latter ordering but not in the former. The only two events that have a place in both orderings are *F* and *A*, which occur in a reversed order in the two orderings. If you put the orderings together, you have a circle. No circle, of course, is an ordering, which becomes clear if one realizes that *A* can be seen both as a soft fact and as a hard fact about the past from any other point of the circle (the same can be said of every point of the circle: in the circle, any point is at once past and future relative to any other point).

When there is anything like such a temporal and causal circle, our intuitions about which facts are "hard" and which are "soft" become hopelessly confuse. No fact seems to be "absolutely hard" or "absolutely soft", even though they seem to be more or less hard insofar as they are more or less near in the past (the nearer in the past, the harder – just as, the less near in the past they are, the nearer in the future and so the less hard). Even in cases where such a temporal and causal circle occurs, grounding does not seem to become circular.[17] The reason why *F* occurs, indeed, is clearly that Riccardo does *A*, not the reverse, even though *A* is subsequent to *F* in the "common" temporal order. So, *A* follows *F* in the common temporal order while preceding *F* in the order of grounding. Likewise, *S*'s doing an action follows in the temporal order the divine foreknowledge that *S* will do that action, but nevertheless it is the ground of that foreknowledge, in the plain sense in which, for any *p*, it is the fact that *p* that grounds the knowledge that *p*, and not the reverse (nothing can be a fact because someone knows that it is a fact; rather someone can know that it is a fact because it is a fact). So, God knows that *S* will do *A because S* will do *A*, and not the reverse. If this is true, one might give

17 Here and in what follows our notion of grounding is largely a non theoretical, naive one – roughly the notion of what relates, for example, the existence of a conference to the existence of its participants and their actions, or the existence of molecules to the existence of atoms, or the truth maker of a sentence to its truth. Nothing will be said here, then, on the many sophisticated questions that are nowadays widely discussed under the heading of "metaphysical grounding".

a definition of "hard" versus "soft" in terms of grounding, regardless of temporal order, roughly along the following lines.

A fact about the past is strongly hard just in case it is grounded *only* in the past while it is weakly hard just in case it is *ultimately* grounded in the past. (Accordingly, the present fact that at t_3 God will know that at t_1 Paul did Y is weakly hard, for it is ultimately grounded in Paul's having done Y in the past by being directly grounded in a future divine knowledge of that action; but divine past foreknowledge is neither strongly nor weakly hard for it is grounded in future facts, so it is a soft fact). A fact about the past is accidentally necessary just in case it is strongly hard. Interpreted in modal terms, accidental necessity is a relation between worlds, while interpreted in terms of grounding it is an intra-world relation between actual facts: the latter is, at any rate, weaker than the traditional notion of accidental necessity. An action's being within one's power, however, remains a modal notion: Y is within X's power at t just in case there is a world W such that X does Y in W, and the same strongly hard facts occur before t in W and in the actual world.

Summing up: S is free in doing A at t iff it is in S's power to abstain from doing A. And to abstain from doing A at t is in S's power iff A's abstention from doing A at t is (logically, physically, biologically, chemically and so on) compatible with t's past, i.e., with the sum total of the world's slices that are antecedent to t, purified from soft facts. Plantinga treats accidental necessity as a strictly modal notion: a fact about the past is accidentally necessary only in case there is no possible action that is capable of preventing it from taking place.[18] On the other hand, he seems to understand the notion of a hard fact essentially in terms of strict "aboutness": a hard fact is strictly about the past (it is, so to speak, intrinsic to the past). Moreover, he believes that some hard facts are not accidentally necessary insofar as there are some facts that are strictly about the past, yet the possibility of acting so as to prevent them from taking place obtains.[19] We prefer to make of two notions one, by giving up the distinction between hard facts and accidentally necessary facts. Hard facts are nothing over and above accidentally necessary facts, and are conceived of in non-modal terms, as facts that are *actually* grounded in the past.

18 Plantinga (1986), p. 249.

19 Plantinga (1986), p. 254. See, the critiques of Fischer in Fischer *et al.* (2009), p. 261.

4

In order to remain true to Ockham's approach and avoid bending his thought to fit a contemporary framework, one must safeguard three fundamental assumptions: i) ontological parsimony, i.e. the principle that one ought not to posit the actual existence of multiple and superfluous things (*res*)[20] by e.g., turning connotative terms into real beings; ii) the non-modality of temporal relations, according to which the term 'time' itself defines connotation[21]; and iii) the doctrine of divine simplicity, which grounds the other two. Subscribing to divine simplicity means to stick to an integrated consideration of the ontological and the theological level, just as Ockham does; but as our analysis is concerned with providing a critical reconsideration of Plantinga's position, which separates the two levels, we opted for considering them separately too. Neo Ockhamist thinkers, like Plantinga and – from a different standpoint but in a similar methodological perspective – Hasker, seem to focus on only one of the two poles that constitute the relation between facts and acts of foreknowledge that Ockham regards as unavoidable. As a consequence of this approach, they are led to build a kind of "bridge relationships" or trans-world relationships between possible worlds, such that the ontic relatum (located in a world w_1, w_2, w_3 and so on, according to necessity) can make the propositions of divine foreknowledge true. But this makes their solutions appear as dangerously *ad hoc*, that is, built just in order to justify the apodictic statement about human free will. In contrast with this approach, and in line with Ockham's principles, we propose to merge the two notions into one, by treating hard facts on a par with accidentally necessary facts. Hard facts are always accidentally necessary, and are defined in terms of grounding, in a non-modal way (so that only the actual world is relevant in order to establish whether some facts are hard or not). We argue that, thus understood, Ockham's way-out is by and large more viable than in Plantinga's modal interpretation.

Indeed, we point out that, in contrast with Plantinga, Ockham postulates a relationship of *bringing about* which occurs completely *inside this* single world, which the principle of ontological parsimony and the doctrinal grounding of faith conceive of as unique. This relationship does not occur across a number of max-

20 The principle of parsimony is strictly connected to the name of Ockham. Even if the idea can be traced to as far back as Aristotle, passing through Roger Bacon, this label perfectly fits Ockham's thought since "his metaphysical conclusions are what we would expect from a philosopher who is assuming this principle as his guide" (Adams (1987), I, p. 157). For an externalist reading of Ockham, see also, Brower-Toland (2007).

21 Guillelmus de Ockham (1985), p. 47.

imal sums of states of affairs (i.e. possible worlds) but across different planes within the same world – which, incidentally, guarantees a more solid grounding than that granted by the unstable range of possible worlds. Thus construed, the relationship does not need a plurality of worlds; it simply moves along a single axis, where the intensity of the entailment obtaining between events is variable, depending on whether it is the ontic or the epistemic plane that is taken into account. Once the principle by which an individual *S* has the possibility to act without her actions being already determined is affirmed, the free act *E* and the free act non-*E* determine which relation is brought about by the implication/capacity: the one with proposition *Y* or the one with proposition *X*. In this way, the incompatibility between the freedom of human will and the necessitating foreknowledge of God is solved by moving them both into different worlds (states of affairs), in a disposition, as it were, which links the divine act of knowledge with the corresponding factual event; such a link does not constitute a simple logical possibility but rather a possible and existing state of affairs (even in the peculiar form of a possible world). This approach enabled Ockham to account for those propositions (e.g. future contingents, prophecies) that are neither true nor false as far as the reference is concerned, given that they rest on future states of affairs that are not yet closed. Such propositions, however, are determinately true according to a truth-value that rests on the premises confirmed by *logica fidei* (i.e. an epistemic logic of religious belief) and not on the accidental necessity of the past. In Ockham's view, theology is the language that provides the *viator* (i.e the man walking this path of life) with the pragmatic rules for connecting *res* on the plane of ontological grounding. Those rules, nonetheless, are not to be understood as actual components of reality, insofar as they are mere connotative terms by which human language is able to understand the relationships that occur between actual things.

The interest in recovering Ockham's original solution, therefore, does not lie with the requirements of philological strictness or historical accuracy; rather, it is due to the fact that Ockham's theory provides a clear and efficient tool for discussing the problems of theological compatibilism: namely, the notion of a unique world, the uniqueness of which guarantees the grounding's necessity and in which events must be conceived of as poles in a relationship of variable intensity. It is in this area of Ockham's thought that one may find an answer to the still open question concerning whether Ockham's way-out is, in any sense, available to non-eternalists too.

Bibliography

Ancient and medieval sources

Augustine of Hippo (1887), "Sancti Aurelii Augustini opera", in *Corpus Christianorum Ecclesiasticorum Latinorum*, Vindobonae: Hoelder-Pichler-Tempsky. (op. cit.: *De civitate Dei libri viginti duos*, PL 42, CSEL 40/1–2; *De libero arbitrio libri tres*, PL 32, CSEL 74; *Retractationum libri duo*, PL 32, CSEL 36).

Anselm of Canterbury (1946–1961) *Sancti Anselmi Cantuariensis Archiepiscopi "Opera omnia" ad fidem codicum recensuit* F.S. Schmitt O.S.B., 6 voll, Edinburgi apud Th. Nelson et filios.

Aristotle (1949), *Categoriae et liber de Interpretatione*, edited by L. Minio-Paluello, Oxford. See also: *Aristotle's Categories and De Interpretatione*, edited by John L. Ackrill, Oxford: Clarendon Press.

Aristoteles Latinus (1968), *Aristoteles Latinus*, IV. 1–4, edited by Lorenzo Minio-Paluello et Bernard G. Dod, Bruges-Paris.

Boethius (1880), *Anicius Manlius Severinus Boethius, Commentarii in librum Aristotelis Peri Hermeneias pars posterior*, Hrsg. Carl Meiser, Leipzig: Teubner.

Boethius (1906), *Commentary on Isagoge*, IIa, Hrsg. Samuel Brandt, (Corpus Scriptorum Ecclesiasticorum Latinorum 48), Vienna-Leipzig: Tempsky and Freytag.

Boethius (2000), *De Consolatione Philosophiae, Opuscula Philosophica*, Hrsg. C. Moreschini, Munich-Leipzig: Teubner. Also: Altra ediz-consultata *Anicii Manlii Severini Boethii Philosophiae Consolatio*, a cura di Ludovicus Bieler, Turnhout: Brepols.

Guillelmus de Ockham (1974), Guillelmi de Ockham "Summa logicae" in *Opera Philosophica I*, edited by Ph. Boehner, New York: St. Bonaventure University.

Guillelmus de Ockham (1978a), Guillelmi de Ockham *Tractatus de praedestinatione et de praescientia Dei respectu futurorum contingentium*, edited by Ph. Boehner, New York: St. Bonaventure University.

Guillelmus de Ockham (1978b), Guillelmi de Ockham "Expositio in librum *Perihermeneias* Aristotelis", in Guillelmi de Ockham *Opera Philosophica*, II/2, edited by A. Gambatese and S. Brown, New York: St. Bonaventure University.

Guillelmus de Ockham (1980), Guillelmi de Ockham "Quodlibeta septem", in *Opera Theologica IX*, New York: St. Bonaventure University.

Guillelmus de Ockham (1985), Guillelmi de Ockham "Expositio in libros physicorum", in *Opera Philosophica IV*, edited by V. Richter and G. Leibold, New York: St. Bonaventure University.

Guillelmus de Sherwood (1983), *William of Sherwood, Introductiones in logicam*, edited by Ch. H. Lohr, P. Kunze and B. Mussler, 219–99. Also in *Traditio*: XXXIX, 219–289.

Scotus, I. D. (1963), Ioannis Duns Scoti *Opera Omnia*, studio et cura Commissionis Scotisticae, Vol. V, Praeside P. Carolo Balić, Civitas Vaticana.

Scotus, I. D. (1966), Ioannis Duns Scoti *Opera Omnia*, studio et cura Commissionis Scotisticae, Vol. XVII, Praeside P. Carolo Balić, Civitas Vaticana.

Contemporary sources

Adams, M. M. (1967), “Is the Existence of God a “Hard” Fact?”, in *The Philosophical Review*: 76, 92–503.

Adams, M. M. (1987), *William Ockham*, Notre Dame, Indiana, (USA): University of Notre Dame Press.

Alston, W. (1986), “Does God Have Beliefs?”, in *Religious Studies*: 22, 287–306.

Brower-Toland, S. (2007), “Intuition, Externalism, and Direct Reference in Ockham”, in *History of Philosophy Quarterly*: 24, 317–336

Craig, W. L. (1986), “Temporal Necessity: Hard Facts/Soft Facts”, in *Philosophy of Religion*: 20, 65–91.

Craig, W. L. (1988), *The Problem of Divine Foreknowledge and Future Contingents from Aristotle to Suarez*, Leiden: E. J. Brill.

Craig, W. L. (1991), *Divine Foreknowledge and Human Freedom. The Coherence of Theism: Omniscience*, Leiden: E. J. Brill.

Fine, K. (2005), “Modality and tense”, in Kit Fine, *Philosophical Papers*, Oxford (UK): Oxford University Press.

Fedriga, R. (2015), in *La sesta prosa, discussioni medievali su prescienza, libertà e contingenza*, Milano: Mimesis.

Fischer, J. M. (1983), “Freedom and Foreknowledge”, in *The Philosophical Review*: 92, 67–79.

Fischer, J. M. (2011), “Foreknowledge, Freedom, and the Fixity of the Past”, in *Philosophia*: 39, 461–474.

Fischer, J. M., Kane, R., Pereboom, D. and Vargas M. (2007), *Four Views on Free Will*, Oxford: Blackwell.

Fischer, J. M., Todd, P. and Tognazzini, N. (2009), “Re-Reading Nelson Pike’s ’Divine Omniscience and Voluntary Action”’, in *Philosophical Papers*: 38(2), 247–270.

Freddoso, A. J. (1983), “Accidental Necessity and Logical Determinism”, in *Journal of Philosophy*: 80, 257–278.

Hasker, W. (2001), “The Foreknowledge Conundrum”, in *International Journal for Philosophy of Religion* : 50, 97–114.

Helm, P. (2010), “Eternity”, in *Stanford Encyclopedia of Philosophy*, edited by E. N. Zalta. URL = http://plato.stanford.edu/entries/eternity/> FALL 2015

Kane, R. (ed.) (2002), The Oxford Handbook of Free Will, Oxford-New York: Oxford University Press.

Knuuttila, S. (2006), “Medieval Theories of Future Contingents”, in *Stanford Encyclopedia of Philosophy*, edited by E. N. Zalta. URL = <http://plato.stanford.edu/entries/medieval-futcont/> FALL 2015

Knuuttila, S. (2013), “Medieval Theories of Modality”, in *Stanford Encyclopedia of Philosophy*, edited by E. N. Zalta. URL= http://plato.stanford.edu/entries/modality-medieval/>FALL 2015

Kretzmann, N. and Stump, E. (1981), “Eternity”, in *Journal of Philosophy*: 79, 429–458.

Maierù, A. (1972), *Terminologia logica della Tarda Scolastica*, Roma: Edizione dell’Ateneo.

Marenbon, J. (2005), *Le temps, l’eternité et la prescience de Boèce à Thomas d’Aquin*, Paris: Vrin.

Marenbon, J. (2013), “Divine Prescience and Contingency in Boethius’s Consolation of Philosophy”, in *Rivista di storia della filosofia*: 1, 9–23.

Michon, C. (2002), “Omniscience, liberté humaine et immutabilité divine: remarques à partir du Tractatus de Guillaume d’Ockham”, in *Boulnois*, Schmutz e Solère, 149–169.
Michon, C. (2004), *Prescience et liberté, essai de théologie philosophique sur la Providence*, Paris: PUF.
Mugnai, M. (2013), *Possibile/Necessario*, Bologna: Il Mulino.
Normore, C. G. (1985), “Divine Omniscience, Omnipotence and Future Contingents: an Overview”, in Rudavsky (1985), 3–22.
Ockham, (1969), *William Ockham: Predestination, God’s Foreknowledge and Future Contingents*, M. M. Adams and N. Kretzmann (eds.), New York: Appleton Century Crofts.
Pinnock, C., Rice, R., Sanders, J., Hasker, W., Basinger, D. (1994), *The Openness of God. A Biblical Challenge to the Traditional Understanding of God*, Downers Grove, Illinois (USA): Inter Varsity Press, & Carslile, Cumbria (UK): The Paternoster Press.
Pike, N. (1965), “Divine Omniscience and Voluntary Action”, in *The Philosophical Review*: LXXIV, 27–46.
Pike, N. (1966), “Of God and Freedom: a Rejoinder”, in *The Philosophical Review*: LXXV(3), 369–379.
Plantinga, A. (1974), *The Nature of Necessity*, Oxford: Clarendon Press.
Plantinga, A. (1986), “On Ockham’s Way Out”, in *Faith and Philosophy*: 3(3), 235–269.
Rice, H. (2010), “Fatalism”, in *Stanford Encyclopedia of Philosophy*, edited by E. N. Zalta. URL = http://plato.stanford.edu/entries/fatalism/FALL 2015.
Rudavsky, T. (a cura di) (1985), *Divine Omniscience and Omnipotence in Medieval Philosophy. Islamic, Jewish and Christian Perspectives*, Dordrecht-Boston: Reidel.
Sanders, J. T. (1966), “Of God and Freedom”, in *Philosophical Review*: 75, 219–225.
Zagzebski, L. T. (1991), *The Dilemma of Freedom and Foreknowledge*, Oxford/New York: Oxford University Press
Zagzebski, L. T. (2011), “Foreknowledge and Free Will”, in *Stanford Encyclopedia of Philosophy*, edited by E. N. Zalta. URL = http://plato.stanford.edu/entries/free-will-foreknowledge/FALL 2015.

Paul Clavier

The Importance of Being Timeless

Eternity is very long, especially toward the end.

A great amount of literature has been devoted to the relationship of God to time. Can we conceive of a timeless personal being? Or even of a timeless being at all? And, granted that the concept of a timeless being be free of contradiction, is it compatible with other attributes or attitudes traditionally ascribed to God, like omniscience, interactive dialogue with created beings, providence, foreknowledge? If, as Nelson Pike suggests, timelessness is not to be deduced from other divine attributes, on which grounds are we to discuss whether it may be ascribed to God?

Another concern is the great variety of views about the nature of time, not only in philosophical debates, but even in everyday's worldview. Because of this variety, it is not easy to get an univocal description of "timelessness". I will assume the following: ascribing timelessness to God would not be relevant if we were adopting the thesis of "unreality of time". If time does not exist, if there are no truly temporal relations nor properties, being timeless amounts to a tautological predicate. In order to be relevant, God's timelessness has to be contrasted with truly temporal modes of existence. Let us grant that some beings, and in particular some agents are – at least partly – temporal beings. Let us assume that there is the A-series of time, that we experience as a passage of the time, consisting in past, present and future instants relatively to the frame of reference of a temporal observer. Let us grant that tensed propositions describe something very important about the world. I will first run through arguments pro and contra divine timelessness. I will then inquire into what a timeless God is supposed to miss. By the way, I will shortly consider some unexpected advantages of divine timelessness as regards the problem of theodicy. I will then focus on the traditional problem of compatibility of God's omniscience and libertarian freedom, and consider to which extent God's timelessness can solve the riddle. I will finally inquire into whether a timeless sovereign God undermines the libertarian freedom of creatures.

https://doi.org/9783110594164-002

1 Eighteen short arguments for and against divine timelessness

Let us first spell out some motives of denying or of ascribing timelessness to God. One reason for preferring a temporal God is the worry of keeping the highest possible amount of analogy between our current concepts and the concepts involved in various versions of theism. This can be expressed through the following argument:

1.

(1) Timelessness does not make sense for us.
(2) A definition of God has to make sense for us.
(3) Therefore, God must not be defined as a timeless being.

Surely, the conclusion does not necessarily obtain. For it could be objecteded that God might be timeless, and nevertheless be defined in terms that make sense for us: for not every predicate convenient to *X* is a definition of *X*. Unfortunately, this reply forgets that in the case of God, there is the claim that God IS all his essential properties at once.

God's timelessness is often supposed to weaken the prior probability of theism. To be sure, framing the idea of a timeless person is a much stranger hypothesis than that of a person in time. But, following Brian Leftow[1], we could try to emphasize that, to some extent, significant parts of our moral experiences are timeless: our intuitions about the value of goodness, when we realize that such or such choice is timelessly valuable, some experiences of forgiveness and atonement (whereas some past wrongdoing is conserved but forgiven), some experience of hope (whereas we are already certain of some future happening)…Cf., *Epistle to the Hebrews* 11, 2: "faith is the assurance of what is hoped for"[2], meaning that faith is assured that what is hoped for will become a reality, experiencing simultaneously a state of affairs as "not yet" AND the same as "already" obtained.

Another kind of argument rests on the alleged likeliness of the creatures to the Creator:

1 Leftow (1991), ch. 3.
2 See, Schreiner (2015), pp. 339 ff.

2.

(1) The creature bears the character of the Creator
(for «every agent is found to produce effects which resemble it»).
(2) The creature is temporal.
(3) Therefore, God is temporal.

But of course, this supposes that all the characters of the creature are to be found as such in the creator, which is obviously false, for it is clear that createdness, for instance, is not to be found at all in the creator. More formally, if x bears some likeliness to y, if x resembles y, it does not follow that they must share all their properties. Defining the similarity does not amount to: x resembles $y \leftrightarrow_{def} \forall F(Fx \rightarrow Fy)$; it is enough that under some description, x resembles y. It amounts only to: x resembles $y \leftrightarrow_{def} \exists F(Fx \& Fy)$.

Anyway, on the other hand, as emphasized by Nelson Pike, "this doctrine (of timelessness) is the ultimate expression of God's transcendence"[3]. Instead of undermining a philosophical construction of theism, it would be a good warrant for it. Following Aquinas *via remotionis*, we have to establish that "we reach whatever understanding we have of God's attributes, by removing 'imperfections' that attend these qualities when possessed by finite things" [4]. But this of course may lead to a remote God. So let us consider other arguments to God's remote perfection:

3.

(1) Being in time is an imperfection.
(2) Timelessness is a perfection.
(3) God is endowed with every perfection.
(4) Therefore, God is timeless.

This is not very cogent, for premiss (1) is highly questionable. Why should not "being in time" be another perfection? To this extent, a godhead endowed with every perfection would have to be both timeless and in time (enjoy temporal and atemporal dispositions). Another argument runs as follows:

3 Cf., Pike (1970), Preface p. xi. Obviously enough, there are no direct arguments to God's timelessness from other attributes. He emphasizes that timelessness entails immutability, incorruptibility and immortality, but not conversely (idem, p. 43). Nevertheless it could be suggested that timelessness be the simplest way (or the best explanation) for being incorruptible, immortal, immutable. Timelessness makes all these attributes void of their original meaning: a timeless being cannot die, but he does not live a life which could cease anyway, a timeless being is incorruptible, for he has not even the time to be corrupted, it is immutable too, but when could he have changed anyway? He does not even keep the same as before... he just timelessly is.

4 Pike (1970), p. 2.

4.

(1) God transcends the creatures.
(2) All creatures are temporal.
(3) Therefore, God is not temporal.

Of course it is a weak argument, for *x* transcends *y* does not imply, but just favours the view that if *y* is an *F*, then *x* is not an *F*. We are in a symmetric case of the relationship of similarity. And once again, the relationship of transcending does not amount to: x transcends $y \leftrightarrow_{def} \forall F(Fy \rightarrow \neg Fx)$; but only to: x transcends $y \leftrightarrow_{def} \exists F(Fx \& \neg Fy)$, with *F* designating a distinctive characteristic of *x*, on a metaphysically significant level, which can hardly be described better than by saying that it is the kind of characteristic exemplified by timelessness.

It would be better, but more begging the question, to say:

5.

(1) Existing in time, successively, is the exclusive property of created beings.
(2) God is not a created being.
(3) Therefore, God is timeless.

There are further arguments to timelessness, concerning the ontological status of time.

6.

(1) Time does not exist apart from the world (time is a relational property or a dimension that does not exist apart from a created world).
(2) God exists apart from the world.
(3) God exists apart from time.
(4) Therefore, God is timeless.

Or, to put it otherwise:

7.

(1) Time supervenes on events (as their relational property).
(2) Events occur to substances.
(3) Substances depend on God the creator.
(4) Time depends on God the creator.
(5) Therefore, God is not submitted to time.

In this view, God creates everything, including the temporal aspects of what there is.[5] God may create at a time, (meaning that the effect of his creative operation may occur at a time) but God is not in time. Although he is operating outside time, God's action has consequences in time.

8.

(1) God possesses completely illimitable life all at once.
(2) Possessing illimitable life all at once excludes temporal extension (vs. once upon a time, or sometimes, or many times or always).
(3) Therefore, God is outside time.

But here the first premiss, inspired by Boethius' definition of eternity, is begging the question, except if the clause "all at once" is revised, so that timelessness is replaced by everlastingness, infinitely extended duration.

9.

(1) God is causally prior to any appearance of time.
(2) Being causally prior to *x* entails being outside *x*.
(3) Therefore, God is outside time.

But it might be objected that every cause is supposed to precede chronologically its effects. We need a further reason. Maybe is it provided by Augustine (*Confessions* XI. xiii (16)), according to whom "It is not in time that you precede times. Otherwise you would not precede all times. In the sublimity of an eternity which is always in the present, you are before all things past and transcend all things future, because they are still to come." But is this claim an argument? Let us try to construe Augustine's justification:

10.

(1) *x* precedes *y* in time $\leftrightarrow_{def}$ *x* exists at a time prior to *y* (*x* and *y* are part of the series of time).
(2) If God preceded all times in some time, there would be a time that God does not precede. That is, a time to which he, (or at least his operating) belongs!
(3) God precedes all times.
(4) There is no time in which God precede all times.
(5) Therefore, God timelessly precedes all times.

5 See, Plato: "Time was created along with the heaven" (*Timaeus*, 38b5), and Augustine: "What times existed which were not brought into being by you? Or how could they pass if they never had existence? Since, therefore, you are the cause of all times, if any time existed before you made heaven and earth, how can anyone say that you abstained from working?" (Augustine, *Confessions*, XI, xiii (15)).

But this looks more like a stipulation than like an argument. In the preceding argument, it has been alluded to God's operation. This suggests the next argument to divine timelessness:

11.

(1) God operates timelessly.
(2) The operation of any being follows its nature (*operari sequitur esse*).
(3) Therefore, God is timeless.

But of course, the two premises are not easily granted. Nevertheless I will try to advocate them in due time (if I may say in this context) hereafter. There is, too, a straightforward argument from simplicity:

12.

(1) If *S* is simple, *S* has no temporal parts.
(2) If *S* has no temporal parts, *S* is timeless.
(3) God is simple.
(5) Therefore, God is timeless.

Some other lines of reasoning to a temporal God may sound more compelling. They emphasize the lack of relationship of a timeless God with his creatures, and suggest that a necessary condition for doing God's job is to share the temporal dimension of the creatural world:

13.

(1) God interacts with temporal agents.
(2) Interaction with temporal agents cannot be but in time.
(3) God interacts in time.
(4) Actions conform to the nature of agents.
(5) Therefore, God is in time.

14.

(1) If God is timeless, nothing really happens to him.
(2) If nothing happens to some person, she is not a living person.
(3) God is timeless, God is not a living person.
(4) God is a living person.
(5) Therefore, God is not timeless.

15.

(1) God intervenes in human history
(he addresses to manhood, gives commands, answers prayers).
(2) The only way to intervene in *X*'s history is to share *X*'s experience of time.
(3) Therefore, God shares human experience of time.

But, in order to reach the conclusion that God is in time, we need at least 2 additional premises:

- That human experience is never timeless.
- And, that God cannot be timeless and share (at the same time!) the human way of experiencing time.

I will turn back later to the issue as to whether a timeless God really misses something.

Some other arguments are taken from God's alleged omniscience:[6]

16.

(1) God knows every event that is logically knowable, including contingent future actions of the free creatures.
(2) All times are equally present to God's knowledge.
(3) Therefore, God is outside time (God is not subject to the *A*-series but has access to the *B*-series, through a non-temporal simultaneity).

But of course, it could be objected that the second premiss is begging the question. And alternative views of what knowing some event amounts to may give the following argument:

17.

(1) God is omniscient.
(2) God knows every event.
(3) Temporal events exist in time.
(4) A true cognizer of temporal events must herself exist in time.
(5) God exists in time.
(6) Therefore,God is not timeless.

6 "All are open and laid bare to his eyes, even those things which are yet to come into existence through the free action of creatures" (Vatican Council I, Dei Filius I: DS 3003; cf., Wis 8:1; Heb 4:13). See also, Calvin, according to whom attributing foreknowledge to God means "that all things have been and perpetually remain before his eyes, so that to his knowledge nothing is future or past, but all things are present; and present in such a manner, that he does not merely conceive of them from ideas formed in his mind, as things remembered by us appear to our minds, but he holds and sees them as if actually placed before him" (*Institutes of the Christian Religion*, Book III, ch. 21, transl. John Allen (Philadelphia, 1813) II, 145).

To put it otherwise:

18.

(1) God is timeless.
(2) A true cognizer of temporal events must herself exist in time.
(3) God does not know every event.
(4) Therefore,God is not omniscient.

Here, as quite often in metaphysics, the *modus ponens* of the one is the *modus tollens* of the other. When knowledge is supposed to involve temporal simultaneity of the cognizant and what is known, omniscience excludes timelessness.

There are also considerations from scriptural suggestions or religious experiences. Not only the concept of a timeless being has to be intrinsically coherent, but it still has to fit some requirements when applied to God. The list of these requirements depends largely on the variety of denominational creeds.

Think, for instance, of the afterlife which is proposed to humans by God. Is it temporal? Then it is filled with a succession of instants. It would be then relevant to claim: "Eternity is very long, especially toward the end". But there may be alternative views of afterlife. Does it really make sense to ask: how much time did you spend in Purgatory? How long will you stay in Paradise? Or John Paul the second has just arrived in heavens, he's not used to eternal life yet…Sure, it is hard to deal with timelessness. My suggestion would be, if there is an afterlife, it is likely to be a life without after. Not a second half or a bonus game.

In some confessions, God's operations seem to escape the temporal frame of reference. Let us give an example: The Blessed Virgin is believed by the Catholics to have been preserved from original sin by the anticipating grace of the Holy Spirit, in view of the merits of Christ, the Redeemer"[7]. This very strange timing is hard for a temporal being. A timeless God is more likely to manage that. There is also the issue as to whether creation is to be conceived of in terms of temporal change.

7 "We declare, pronounce, and define that the doctrine which holds that the most Blessed Virgin Mary, in the first instance of her conception, by a singular grace and privilege granted by Almighty God, in view of the merits of Jesus Christ, the Saviour of the human race, was preserved free from all stain of original sin, is a doctrine revealed by God and therefore to be believed firmly and constantly by all the faithful." (*Ineffabilis Deus*, Pius IX).

2 What is a timeless God really missing?

Let us now turn to consider what is a timeless G od really missing. As often emphasized (Coburn, Kenny), God's timeless existence is supposed to be inconsistent with God's being a personal agent, genuinely interacting with human beings,which seems to be a basic requirement for an suitable concept of godhead.[8] Let us imagine the following dialogue between a poor pious beggar and God:

> O Lord, What are for you Centuries? Nothing but minutes! What are for you billions? Nothing but a few cents. So please, God, give me a few cents!
> Okay, beggar, wait a minute!

In this dialogue, there is a misunderstanding as regards the urgency of the petitionary prayer. The petitioner needs the cash now. And God's extratemporal answer is, supposedly, not fitting: and this unappropriateness is the *nervus ridendi* of the joke.

Seemingly, addressing some person in a true dialogue supposes that we can share with her the meaning of temporal indexicals ("now", "then", "yesterday", "from now on", "I will be with you till the end of the world", "Today you will be with me in my paradise", and so on). This requires that we have a common understanding of these chronological indexicals. The main objection to God's timelessness is then that, if God is timeless, he does not experience the coming to be nor the passing-away of things, actions and events. He can make no difference between what we call past, present, future. He has no true knowledge of temporal realities.

8 "If we are to characterize God at all, we must say that He is personal, and if personal then temporal, and if temporal then in some sense in time, not outside it" (Lucas (1989), p. 213). Cf., G. Jantzen (1983), p. ...: "A timeless and immutable God could not be personal, because he could not create or respond, perceive or act, think, remember, or do any of the other things persons do which require time. Thus, within the framework of a theology of a personal God, the doctrines of divine timelessness and immutability cannot be retained".
Well, being a personal agent does not entail necessarily being in time, although our common pattern of personal agency is associated with temporal decision. But despite the temporal feature of our moral experience, the very responsibility cannot be said to be essentially temporal. Essential to a personal agent is his causal responsibility for some deeds or operations. I suggest that the relationship between my choice or my decision and their object is not essentially temporal, even if some consequences of my decision may occur in time. First, I may discover ex post that what I have experienced is what I deeply considered as right or wrong. My moral agreement or disagreement has not forcedly to precede the experience in order to be relevant.

This is the bedrock of some arguments against God's timelessness:[9]

19.

(1) In a genuine relationship, the related persons have to interact.
(2) Interaction requires intervals of successive instants common to both related persons.
(3) A timeless being does not exist at successive instants, it only exists.
(4) God's timelessness precludes his partaking a succession of common instants.
(even if some kind of duration can be ascribed to a timeless being, it is "an infinitely extended, pastless, futureless duration").
(5) God's timelessness precludes genuine interaction with timely persons.
(6) God genuinely interacts with timely persons.
(7) Therefore, God is not timeless.

This requirement of successive instants in common looks more obvious when the interaction is conceived of in the manner of a dialogue:

> In a dialogue, the interlocutors have to alternate answers and queries.
> This alternation implies both interlocutors experiencing the same or at last overlapping intervals of time (common segments of duration).
> If one of the interlocutors is timeless, he does not really dialogues with the other.

As N. Wolterstorff puts it:

> Some of God's actions must be understood as a response to the free actions of human beings – that what God does he sometimes does in response to what some human being does. I think this is in fact the case. And I think it follows, given that all human actions are temporal, that those actions of God which are "response" actions are temporal as well. (Wolterstorff (1975), p. 197.)

Of course both of interlocutors must share something. The questions raised must have, if not the same, at least some common meaning to the creator and the creature. But is it required that they must share the same experience of the passage of time? I do not think so. For even human interlocutors are never ensured they share the same experience of time.

But, it must be recalled once more, that the timeless view of God's knowledge is not that of a preview, that God's knowledge of the *B*-series of time *does not* consist in a foreknowledge or in some infallible forecast.

I would like to mention a famous instance of interactive dialogue with an absolutely timeless immutable God[10]. All along the narrative and even in the specu-

9 Cf., Stump and Kretzmann (1981), p. 45.

10 "In you it is not one thing to be and another to live: the supreme degree of being and the supreme degree of life are one and the same thing. You are being in a supreme degree and are

lative sections of the Confessions, it doesn't seem that Augustine is troubled with God's immutability, which in no way precludes pathetic episodes of interaction, like the famous narrative of his conversion. The famous *Tolle lege*, a childish song encouraging Augustine to read a passage of the scriptures, and the passage of the scripture, are experienced as warnings and answers to Augustine (and Monique) distress and prayer. If God timelessly brings about these circumstances, he is not to be compared any more with the remote organizers of a treasure hunt, preparing in advance for unknown future generations riddles, trials traps and rewards. His answers and advices arrive in due time, with equal or more relevance than any attentive answer in a temporally developed dialogue. A timeless God is not forcedly an anonymous prerecorded speech-server. (Like: Please hold the line, if your query is about the meeting in Warsaw, please dial 1)

Well and good, but there is still the objection as to whether God really experiences our temporal condition. God is supposedly omniscient. An omniscient being must know what it is like to experience the passage of time. Does it mean that he has to be acquainted with the passage of time? Open theists would insist: of course he has to. Eternalists would claim that it is possible to know what it is like to experience the passage of time without having to experience it.

We need further considerations on the knowledge of a timeless God. There are at least three models of God's knowledge of events: predictive, observational, practical. The predictive model is missing the point of free libertarian actions, unless we accept compatibilism, and anyway is not that of timeless God. The observational model seems to imply that God must be somehow present when a future event occurs.

Surely, this epistemic access to future events is possible on Stump and Kretzmann account of ET simultaneity. But we may need an additional support for the view of an epistemic immediacy, according to which the timeless God can directly and eternally be aware of the temporal events without having to attend temporally the scene.[11]

This support is God's "practical knowledge". God has a practical knowledge of every singular beings and modes, since he is the one who generates and sustains them. He knows everything that happens by bringing about the very existence and operation of every object involved in every state of affairs, including

immutable". (Augustine *Confessions*, I. vi (10)).

11 This is the challenge suggested by T. D. Senor (2009), p. 85. See also, p. 74, where the challenge of a non temporal epistemic presence to temporal events is nicely defined.

free deeds and demands made by some creatures.[12]

If we deny this view, then we get a dualistic view, according to which there is or there are objects and operations which do not depend on God. God knows what the world is like, and may even know what it is like to be a bat, a vat, a brain in vat, a heroe, a villain, an ordinary person, and so on. In order to do so, he does not need to experience perspectives on the world. In order to be truly omniscient, God must not have all the phenomenal concepts (concepts about what it is like to have such and such phenomenal experiences any finite conscious beings may have). God does not need nor want any point of view. He sees ("And God saw..." that is: He sees timelessly) what he makes.

Kretzmann once compared the knowledge an omniscient being has of the entire scheme of contingent events with "the knowledge you might have of a movie you had written, directed, produced, starred in and seen a thousand times. You would know its every scene in flawless detail, and you would have the length of each scene and the sequence of scenes perfectly in mind." Nevertheless, on Kretzmann's account, there was an objection as to whether the movie-one-man-band was really omniscient: "You know the movie immeasurably better than do the people in the theatre who are now seeing it for the first time, but they know one big thing about it you don't know, namely, what is now going on on the screen. Thus, the similar account of omniscience regarding contingent events is drastically incomplete. An omniscient being must know not only the entire scheme of contingent events from beginning to end at once, but also *at which stage of realization that scheme is now.*" [13]

But this lack of indexicality, as already emphasized, is not a lack of knowledge. The questions: where am I? and what time is it now? can be answered by a timeless being, who knows exactly the order and series of spatial and temporal locations of every being. God the generating and sustaining cause of the universe knows that because He makes that. It could be said that the timeless being, far from missing the "now", knows all the "nows".[14]

---

12 T. Flint (1988), p. 35, precludes the account that God knows our free actions by knowing his own intentions to cause us to act in certain ways, for such external causation is, according to the libertarian, is incompatible with freedom.

13 Kretzmann (1966), p. 414.

14 Inquiring into what kinds of things a timeless being know, wondering whether there is a timeless knowledge of what is happening now, Nelson Pike denies that statements such as "the first scene is now on the screen" and "Today is the twelfth of May" report facts that are unknown from a timeless being, for either these are not facts, or they could be referred to through statements free of temporal indexical expressions. As he puts it "all that has been established is that there are certain *forms of words* that a timeless individual could not use when formulating or reporting

If the omniscient being is the creator, that is the generating and sustaining cause of the universe, he is actually responsible for each stage of realization. In this sense, everything is timelessly present to God, but this neither requires nor entails that God has an experience of everything happening at once. He has a nonsequential knowledge of the *A*-series. In this sense too, God timelessly makes everything happen in due time. As Hasker put it, "The way God knows things to be is the way things really are" does not entail "The way in which God knows things (i.e., his manner of knowing them) is the same as the way in which they exist."[15] Therefore the following view seems to me consistent:

> God timelessly brings about a world made of successive temporal events.
> God knows what he is timelessly doing.
> God is not committed to observe within a temporal framework what he otherwise perfectly knows by doing it.
> God knows perfectly what's going on, what happens to everyone, everywhere and always.

He knows, because he does. According to Anscombe, being the first agent timelessly involved in every action, he could say "I do what happens"[16]. (This claim

---

his knowledge". (Pike (1970), p. 95) Katherin A. Rogers complains that on Stump and Kretzmann eternalist view, or on Leftow's "QTE (Quasi Temporal Eternity)", "God's supposed duration has nothing in common with the duration we know, which just is temporal extension." (Rogers (1994), p. 14) Once again, my defence of God's timelessness doesn't lack the temporal extension, for God brings it about. If he does not experience it, he nevertheless makes it. No part of the *A* series is not due to God's generating and sustaining power.

15 Hasker (1989), p. 166.

16 Anscombe (1957/1963), p. 53. Cf., Aquinas ST II–II (q. 33, a.1c): whereas "speculative reason only apprehends things, [...] practical reason not only apprehends but also causes them". The concept of non-observational knowledge is first compared with the knowledge one has of the position and movements of one's limbs, that can be known "even with your eyes shut" (I, 15), and without there being any "separately describable sensations" (ibid., 13) that give rise to your knowledge. "Later on, she compares the knowledge one has of one's actions to the knowledge of a project supervisor who directs the construction of a building from afar, without seeing or hearing any reports on its progress (ibid., 82); to one's ability to know what one is writing even if one's eyes are closed (ibid., 53, 82); to God's knowledge of creation (ibid., 87); and to a list that a shopper carries with him that directs his purchases, in contrast to a list made by a detective who follows the shopper around. Anscombe (1957/1963), p. 56)' (Schwenkler (2015), p. 29). Of course, among the instances of practical knowledge, God's causation of the world is the only timeless one, if we adopt the ontological relational view, rather than the transitional account of creation. So we have to justify that the former holds. In the transitional account, "once there was nothing (but God), then there was something", creation is supposed to describe the transition from nothingness to being. Something is supposed to happen to nothing, which happens to become something. This amounts to ascribe properties to "nothing", which is absurd.

sounds like a verse from the Old Testament or from the *Baghavad Gita*). I will further inquire into whether this practical omniscience undermines libertarian freewill.

3 The importance of being timeless in theodicy

I suggest to sketch briefly one consequence of timlessness in the problem of theodicy. A temporal God is supposed to share the destiny of the world, whereas a timeless God is, supposedly, remote and indifferent, splendidly isolated. But let us notice that on both views, God is equally responsible for their permission. Nevertheless, if God is temporal, He cannot foreknowledge free libertarian contingent actions (if compatibilism is excluded). He discovers the horrendous evils that are due to free libertarian contingent actions as they go along, day after day. He has no reason to allow so much evil to occur, since He doesn't know if the future will not be worse. Any significant evil would be enough and a good reason to stop the whole process. This view is to some extent unbearable when one turns to consider horrendous evils that God is supposed not only to allow, but actually to sustain. Given the historic accumulation of forfeits of all sorts, why, if God is observing the improvement of evil, does not He stop the process? On the timeless view, the issue is not that of the intervention, but of the timeless permission of evil. And on this point, a timeless good God is more reliable in permitting evils, than a temporal God. The former knows how the whole story turns out, then if He is a good God, he cannot have permitted something that turns bad. The latter cannot know what the free libertarian will of the creatures will produce. Evils occur in spite of him; he has no idea of what can happen due to free libertarian choices of some creatures…But this does not diminish his responsibility, for if God is the generating cause of the world, he remains the chief responsible of everything that occurs in the world, even of what happens through secondary causes. If, being timeless, God knows the consequences of the misuse of freewill, he may have permitted their occurrence, for the sake of greater goods like libertarian freedom, courage, moral responsibilty.

But if, being temporal, God could not predict the consequences of the misuse of freewill, he ought not to have endowed his creatures with this power. As regards the problem of theodicy, the temporal view of God is not in a better position than the timeless view.

4 The importance of being timeless as regard omniscience

Let us now focus on the importance of timelessness as regards the problem of incompatibility between divine omniscience and libertarian freedom. Let us first recall roughly the incompatibility argument of free will and foreknowledge. I suggest a compact rephrasing of the argument:

20.

(1) If at t_0 prior to t, God infallibly believes that S performs a at t, it is unavoidable that S performs a at t.
(2) If it is unavoidable that S performs a at t, S could not have done otherwise.
(3) S freely performs a at t only if S could have done otherwise (PAP).
(4) Therefore, If God infallibly believes that S performs a at t, S does not freely perform a at t.

There is a classical way out, suggested by Ockham and his interprets. The way implies that libertarian freedom consists in a counterfactual power on God's past beliefs. Let us define this power:

(CP/GPB) It was within S's power at t to do something such that if she did it, God would have not held the belief he had at t_0 prior to t

If so, the truth of the proposition describing God's belief depends of some future event (what S will have performed at t). The proposition describing God's belief is said to express a soft fact.

But what, if God's beliefs are timeless? Hugh Rice asserts that divine timelessness is not a good way out for escaping the incompatibility between foreknowledge and libertarian freedom. Rice considers that "the view that God is timeless would not solve the problem posed by the incompatibility argument, because there would be an equally good argument for the fixity of God's timeless belief." (Rice (2006), p. 134) Why? According to Rice, if God timelessly believes that some (free libertarian) agent performs some action a at t, this person must have the power so to act that God ***would have not timelessly believed*** that she would perform a at t, namely by refraining from performing a at t. Let us now define the counterfactual power of a person on God's Timeless Beliefs:

(CP/GTB) It was within S's power at t so to act that God ***would have not timelessly believed*** that she would perform a at t, namely by refraining from performing a at t.

As Rice explains, we're unable to affect the past whereas we're able to affect the future because it is neither actual nor real. And, in his view, the power to act otherwise is linked with a power of affecting the future. But Rice seems to imply that since God's timeless beliefs are not about the future, they are not affectable by our power to act otherwise. If God's timeless beliefs are not about the future, they are hard facts.

But it seems to me there is a flaw, for if God's timeless beliefs are not about the future relatively to a timeless God, it does'nt follow that they are past or fix. They are just timeless. God's timeless belief that John performs freely *a* at *t* does not entail that it is fixed at t_0 prior to *t* that God believes that John will perform freely *a* at *t*. There is no need to make the truth of God's belief depend on a future event. For the event is future relatively to the temporal framework of *S*, but not relatively to God. To put it briefly, there is no future for timeless beliefs. As often emphasized by Eleonore Stump, God's infallible knowledge is not a foreknowledge which undermines the exercise of incompatibilist freewill (I mean freewill with the power to act otherwise). To this extent, as Kretzmann and Stump put it, "the short answer to the question whether God can foreknow contingent events is no". God's atemporal omniscience is in no way a foreknowledge, for his beliefs are not temporal. Let us elaborate more carefully the point:

> God's belief at t_0 prior to *t* that *S* will freely perform *a* at *t* is no doubt a belief about the future.
>
> If *S*, as a non-compatibilist free libertarian agent, is able to affect the future, God's belief about what *S* performs in the future depends on *S*.
>
> But God's timeless belief that *S* will freely perform *a* at *t* is not about the future relatively to God.
>
> Then the conditional "if *S* had done otherwise, God ***would have not timelessly believed*** that she would perform *a* at *t*" does not obtain.[17]

17 To this extent, it could be said that literally, God never believes *p*, meaning that there is no *t* at which God's belief takes place. Of course, it is true at every *t* that God timelessly believes that *p*... So to say, it is always true that God believes *p* timelessly, and not true that God always believes *p*. Following this line of reasoning, we are not committed to deny the principle of the necessity of the past. The beliefs of a timeless God are not bound by this principle. Not that God can bring about any change in his past beliefs. But precisely, a timeless God has no past beliefs. He timelessly has beliefs about events, some of them are past from our temporal frame of reference. And it is pointless to talk any person having the power so to act that God would have not believed that they act this way. For a true timeless belief cannot be revised.

Identifying libertarian freedom with a power to affect the future, and then to make something such that if one did it, God would have not held the belief he had, this may be legitimate, as long as God's belief are temporal. But it is precisely what is at stake. If God's belief are timeless, the fact that they are not affectable by *S* does not make those beliefs fix.

Of course it would be problematic that God's infallible (timeless) knowledge of free actions be revealed before they occur. Suppose God (timelessly) reveals to some prophet at t_0 that John performs *a* at *t*. (This is known as the problem of the possible prophet). In this the case, John's performance at *t* is a either a determinist action, or a case of compatibilism. But, as a matter of fact, does it occur? Do we need at all cost to warn off the spectre of divine timeless beliefs rendered fix by their revelation in some time?

Let us consider briefly some examples. Neither Judas betrayal, nor Peter's denial have to be considered like free libertarian incompatibilist acts. Peter is precisely predicted that he will deny three times, but that when he has recovered his mind, he will be able to strengthen his brothers (Luke, 22: 31–34, and cf., 22: 22: "And truly the Son of man goeth, as it was determined"). And the acceptance of the blessed Virgin could be a case of compatibilist answer (is not she filled with divine grace). The possible prophet is a conceptual problem, but there may be no occurrence, from the part of God, of one's free action been revealed to some person so that she couldn't act otherwise and nevertheless would be supposed to act freely.

So, contrarily to what Rice suggests, God's timelessness does not have the consequences he seems to imply in terms of God's beliefs being fix[18].

Are we then committed to conceive of divine timeless beliefs in terms of soft facts? Are we committed to soft eternity? But how could a timeless belief depend on a temporal fact, like John's performance of *a* at *t*? I will suggest that John's free libertarian performance at *t* depends on God's timeless belief, and not conversely. God's practical knowledge may account for his timeless omniscience.

But of course, as we have seen in considering God's practical knowledge, there is still the issue as to whether it is consistent with libertarian freedom.

---

18 As Rice (2006), p. 139, puts it: "the incompatibility argument would at best provide additional support for one's belief that God is timeless. It would not provide an independent reason for it." I will try to search independent reasons for God's timelessness.

5 Final remark

I have suggested that the radical ontological dependence of the world on God's creating operation accounts for his perfect knowledge of what happens: he knows what he makes. It can be objected that the cost of this view is the lack of autonomy of the creature. My last point will be to argue against this objection. What does the statement "God brings about that *X* plays the piano" amount to? Well, we should explain: God says "Let *X* play the piano" and *X* played the piano. And it was good (suppose *X* = Wladyslsaw Szpilman). Now if you ask whether Szpilman plays the piano himself, you will be answered: yes he does. If you insist, by asking whether he plays alone, the answer is still: yes. None else, at the moment, plays the piano but Szpilman. And nevertheless, ultimately, God makes Szpilman doing it alone. God does not play the piano instead of Szpilman. But God remains the Necessary and Sufficient Condition of Szpilman's playing the piano. God is the primary cause of *X*'s playing the piano[19].

No part of Szpilman's playing the piano is not due to God; but God's making Szpilman play the piano implies that Szpilman, not God, plays the piano. Szpilman may be freely playing the piano, or contrained by a nazi officer, this does not matter.

The kind of operation we ascribe to God when we say he makes Szpilman play the piano is specifically different from the kind of actions or reactions we ascribe to the piano-player. It is not the same issue, and the term "making", "maker" have not the same meaning when applied to a timeless creative and sustaining principle, as when applied to a temporal created and sustained thing. Furthermore, this operation is still utterly different from the action we normally describe with factive verbs. The meaning of the factive verb in the sentence "God makes Szpilman play the piano" is not of the same kind than, say, in the sentence: "Szpilman makes you weep". The difference is not merely the difference between accidentally and essentially ordered causes. Szpilman could be an essentially ordered cause of your

---

19 Cf.,"For it is God who works (who is at work) in you to will and to act in order to fulfill his good purpose." (*Epistle to the Philippians* 2: 13) We advocate here a classical theistic view, following Aquinas: "And thus God may be said to be the cause of an action by both causing and upholding the natural power of the being. He gives everything the power to act, and preserves it in being and applies it to action, and inasmuch as by his power every other power acts. And if we add to this that God is his own power, and that he is in all things not as part of their essence but as upholding them in their being, we shall conclude that he acts in every agent immediately, without prejudice to the action of the will and of nature." (*De Potentia*, 3, 7, corpus). But we intend to emphasize that this view makes the paranoid fear of strong concurrentism needless.

laughing, being continually the one who makes you weep, without being the ultimate agent of your being weeping. The making of Szpilman playing the piano does not occur in this world, contrarily to his (eventually) making you weep (and also to the playing of the piano). The causal dependence upon God does not interfere with the causal contribution of secondary causes. They are not causes of the same kind: the first operates timelessly, the second in time. Talk of "primary" and "secondary" causes is misleading, for it seems to imply that the former differ from the latter only in size and time. But a timeless God is not chronologically prior, or posterior to what happens; he has not to be temporally simultaneous with any state of affairs he would then had to observe. He is the timeless Maker of a temporal world, knowing exactly what he makes. He does not overdetermine nor destroy freedom.

To conclude: it is not easy to assess the balance of the arguments pro and contra God's timelessness. The temporalist view is supposed to provide us with an analogical concept of divine person, whereas divine timelessness makes God lacking an essential feature of godhead such as being a true living person interacting with his creatures in a genuine dialogue.

I have suggested to emphasize the threefold importance of being timeless: 1°. It solves the problem of incompatibility between divine foreknowledge and libertarian freedom better than the temporalist view, by cancelling the concept of foreknowledge; 2°. God's timeless omniscience has not the scandalous consequences in theodicy it is often associated with; and 3°. God's timeless omniscience does not undermine, but underpins the autonomy of creatures.

Bibliography

Anscombe, G. E. M. (1957/1963), *Intention*, Oxford: Basil Blackwell.

Anselm (1998), "The Monologion", translated by S. Harrison, in *Anselm of Canterbury, The Major Works*, edited with an Introduction by B. Davies and G. R. Evans, Oxford: Oxford University Press.

Augustine (1991), *The Confessions*, translated by Henry Chadwick, Oxford: Oxford University Press.

Boethius (1969), *The Consolation of Philosophy*, translated by V. E. Watts, London: Penguin Books.

Burrell, D. (1986), *Knowing the Unknowable God*, Notre Dame: University of Notre Dame Press.

Calvin, J. (1960), *Institutes of the Christian Religion*, kedited by John T. McNeill, translated by F. L. Battles, London: S.C.M..

Craig, W. (1998), "The Tensed vs. Tenseless Theory of Time: A Watershed for the Conception of Divine Eternity", in *Questions of Time and Tense*, edited by R. Le Poidevin, Oxford: Clarendon Press, 221–250.

Craig, W. (2001), *Time and Eternity: Exploring God's Relationship to Time*, Wheaton: Ill. Crossway Books.
Cross, R. (2005), *Duns Scotus on God*, Aldershot: Hants., Ashgate.
DeWeese, G. J. (2004), *God and the Nature of Time*, Aldershot: Hants., Ashgate.
Flint, T. (1988), *Divine Providence: The Molinist Account*, Cornell University Press.
Ganssle, G. E. and Woodruff, D. M., eds., (2002), *God and Time*, Oxford: Oxford University Press.
Halbertal, M. and Margalit, A. (1992), *Idolatry*, Cambridge: Mass. Harvard University Press.
Hasker, W. (1989), *God, Time, and Knowledge*, Ithaca, N.Y.: Cornell University Press.
Helm, P. (1988), *Eternal God*, Oxford: Clarendon Press.
Jantzen, G. (1983) "Time and Timelessness", in *A New Dictionary of Christianity*, edited by A. Richardson and J. Bowden, London: SCM.
Kenny, A. (1979), *The God of the Philosophers*, Oxford: Clarendon Press.
Kretzmann, N. (1966), "Omniscience and Immutability", in *Journal of Philosophy*: 63(14), 409–421.
Leftow, B. (1991), *Time and Eternity*, Ithaca: Cornell University Press.
Lucas, J. R. (1989), *The Future*, Oxford: Blackwell.
McTaggart, J. M. E. (1908), "The Unreality of Time", in *Mind*: 17, 456–473.
Maimonides, M. (1963), *The Guide of the Perplexed*, translated by S. Pines, Chicago: The University of Chicago Press.
Pike, N. (1970), *God and Timelessness*, London: Routledge and Kegan Paul.
Rogers, K. A. (1994), "Eternity Has No Duration", in *Religious Studies*: 30, 1–16.
Rice, H. (2006), "Divine Omniscience, Timelessness and the Power to do Otherwise", in *Religious Studies*: 42, 123–139.
Schreiner, T. R. (2015) *Commentary on Hebrews*, Nashville: B&H Publishing Group.
Schwenkler, J. (2015), "Understanding 'practical knowledge"', in *Philosopher's Imprint*: 15, 1–32.
Senor, T. D. (2009), "The real Presence of an eternal God", in *Metaphysics and God, Essays in Honor of Eleonor Stump*, edited by K. Timpe, New York and London: Taylor and Routledge.
Stump, E. and Kretzmann, N. (1981), "Eternity", in *Journal of Philosophy*: 78(8), 429–458.
Swinburne, R. (1977), *The Coherence of Theism*, Oxford: Clarendon Press.
Swinburne, R. (1994), *The Christian God*, Oxford: Clarendon Press.
Wolterstorff, N. (1975), "God Everlasting", in *God and the Good: Essays in Honor of Henry Stob*, edited by C. Orlebeke and L. Smedes, Grand Rapids, Mich. Eerdmans, 181–203.

Christian Kanzian

Temporal Relations as Epiphenomena

The title of our volume is "God – Time – Infinity". My article is mainly dedicated to the second item, namely: time. My aim here is to sketch a theory of time – or more specifically of temporal relations, since I am primarily interested in McTaggert's B-series.

My approach is ontological, in that my theory about temporal relations is embedded in the context of a specific framework of ontological categories. To make a long story short, I contend that temporal relations are epiphenomena, constituted by events.

In order to make this thesis and its implications plausible, it is necessary to present a framework of categories in which events have their proper place (section 1). I next attempt to clarify how events could be the ontological basis of temporal relations (section 2). My next step is to address what it means to say that temporal relations are constituted by events. A key idea is that constitution is a formal relation which implies the epiphenomenal status of what is constituted (section 3). Finally, I aim to apply my account to God's relation to time and to the relation between infinity and time (section 4).

1 The ontological framework

My presentation of the ontological framework in which I am working necessarily lacks specificity. It is a sketch, but it nevertheless includes all of the categories that I regard as the basic elements of reality.

Thus, I assume that there are *substances* or, as I call them, *things*; I also assume *modes*, and so-called *occurrences*, which include *states* and *events*. Moreover, all of these elements of being are particulars: concrete (i.e., non-abstract) individuals (i.e., non-universals).

The central category of my ontology is that of things, the most important ontological feature of which is *three-dimensionality*. Things have three spatial dimensions but no temporal one. They are not extended in time but only in space: they have no temporal parts, but only spatial ones. This implies that they are wholly present at each moment of their existence. They are in a strict sense identical with themselves not only synchronically, at each moment, but also throughout time (or diachronically), and across possible worlds (or modally). They are *endurers*, in

https://doi.org/9783110594164-003

David Lewis's sense.[1] Things exist independently in a specific and unique sense. This is why they may be called the primary entities. However, they will not be the focus – at least not explicitly – of this article.

As endurers, things are the bearers of properties. Things are not bundles of properties, for their nature does not consist of properties. They are characterized by properties. Moreover, being so characterized belongs to the nature of things.

The properties that characterize things are, ontologically speaking, *modes*. Modes, which are my second category, are individual qualitative "determinates", such as this brown of this table. They are strictly or existentially dependent on the things which they characterize. This brown cannot exist without this table.[2]

Modes will play only a minor part in my theory, but they must be mentioned to convey a sense of the ontological composition of the third category: occurences, which comprise states and events. Modes, in virtue of characterizing things, are parts of the composition of states. This brown, of course together with the table, composes the state of the table's being brown. If along with static modes like colour- or mass-determinates we also admit dynamic modes, which can be analysed as alterations of non-dynamic ones, we can assume that such dynamic modes – together with their bearers – compose entities that are ontologically very close to states: namely, *events*. I assume that states and events are ontologically so similar that we may subsume them under the category of *occurrences*. All occurrences have ontological features that clearly distinguish them from things, most notably that they are spread out in time as well as space. Occurrences must also be distinguished from modes, which I will not discuss here. As I said above, occurrences, particularly events, will play the largest role in the present theoretical enterprise, especially insofar as they serve as the ontological basis of temporal relations.

2 Events as the ontological basis of temporal relations

The key to understanding events as the basis of temporal relations lies in the specific four-dimensionality of events. Both states and events are four-dimensional entities, but we can safely ignore states in this discussion. So the remaining question is how the four-dimensionality of events is best understood.

1 Cf. Lewis (1986), p. 202.

2 I use "mode" in the sense of Lowe (2006) or Heil (2003).

We can start by examining a seemingly uncontroversial statement by Jonathan Lowe, who writes that "time necessarily involves change – by which I mean that time necessarily involves *happenings* or *events*."[3] According to Lowe, events are necessary for time and temporal relations. Without events, there is no time – or, in broader terms, in a static cosmos there are no temporal relations.

If we accept this fundamental thesis, we can plausibly assume that the *direction* of time will also have to do with events and their *internal structure*, where the latter can be understood as the succession of different parts or phases.

earlier later temporal relations

↑

b.	**p.2**	**p.3**	**p.4**	**p.5**	**p.6**	**p.7**	**e.**

event-phases

Events have a beginning (b.), or a first phase, which is succeeded by some other phases (p. 2–7), until the events come to an end in the last phase (e.). Temporal relations, of which the relation earlier/later is paradigmatic, seem to be immediately derived from this succession of event-parts or -phases. The beginning is *earlier than*, and the ending *later than* the middle-parts of an event. It is outside the scope of this article to apply the earlier/later relation to a reconstruction of other time-indicating concepts like "temporal overlapping" or "simultaneity". The present aims are limited to discussing, with reference to this scheme, the way in which the temporal dimension of four-dimensional events manifests itself: events have an inner structure from which temporal relations are immediately derived. Our speaking about the fourth dimension of events refers to no more than this.[4]

In the next section we will take a more granular look at this immediate derivation of temporal relations from the succession of event-phases. Constitution as a formal relation or tie will, as mentioned above, play the decisive role.

3 Constitution: the formal tie between events and temporal relations

This section will show temporal relations to be constituted by events. In order to fulfill this ontological function, constitution must not be understood by itself as a

3 Lowe (1998), p. 121.

4 I must neglect the three spatial dimensions of events, to come to my interpretation of the forth, the temporal one.

dyadic entity. For if it were, it would be something genuinely ontological linking events and temporal relations, which would contradict the postulated *immediate* connection between events and time. Moreover, constitution must be assumed to be a relational tie, which generates a specific *dependence* of temporal relations on events. We must accept this if Lowe's implication is to have an ontological basis. Last but not least, the specific constitution-dependence must be shown to be consistent with the *epiphenomenal status* of temporal relations.

In order to capture these functions of constitution, we can introduce it as a particular kind of formal relation or tie. I do not intend to present a full ontology of formal relations;[5] I will instead restrict myself to the minimum that is necessary to discuss constitution as a formal tie that links that which is constituted with that which does the constituting, by way of an epiphenomena-specific ontological dependence relation.

3.1 Formal relations

We take our first important hint about formal relations from Kevin Mulligan's conception of *internal* or, as he calls them, *thin relations*. According to Mulligan: "…a relation is internal with respect to objects *a*, *b*, *c* etc., just if, given *a*, *b*, *c* etc., the relation must hold between and of these objects".[6] This means that the existence of some objects is sufficient and necessary for the relations in question. Jonathan Lowe, who calls Mulligan's internal relation "grounded relations", regards them as "entirely determined by their relata". They offer "no additions to reality", on Lowe's account.[7] Lowe's "no addition to reality" statement means that relations which are a) necessarily and sufficiently given with the existence of some objects, and b) entirely determined by these objects, are themselves no entities or elements of beings.

There is a wide range of relations that fall under this initial definition of internal relations; one example is the bigger/smaller relation. Another is the relation of characterization, which occurs between a mode, a particular property, and a thing which the mode characterizes. For bigger/smaller this is clear: given an object *x* with size *F*, and an object *y* (not numerically identical with *x*) with size *G* (not qualitatively identical with *F*), the bigger/smaller dyad is completely deter-

5 For more details concerning formal ties see my article "Existential Dependence and other Formal Relations" in: Szatkowski (2015).

6 Mulligan (1998), p. 344.

7 Lowe (2006), p. 46.

mined and therefore grounded in *x* being *F* and *y* being *G*. Applying Mulligan's and Lowe's rule, bigger/smaller makes no contribution to reality as an entity in itself. An analogous point holds for characterization.

The choice of my examples (bigger/smaller, and characterization) should indicate both the scope and the diversity of the relations at issue. This raises the question: Shouldn't we assume different groups among our non-entity-relations?

Lowe presents a criterion for distinguishing different types of grounded relations; his criterion is given by the observation that those relations whose occurrence is due to the *nature* or the ontological *form* of their founding instances may be distinguished from those for which this is not the case.[8] Take characterization for instance. If a mode *F* characterizes an object *x*, it combines with *x* because of what it is (namely, a mode), and it does so due to its nature or form, which consists in being a way in which a thing is. I suggest calling such internal relations *formal*.

Take as a contrary example bigger/smaller. That an object *x* is bigger than an object *y* has nothing to do with its nature or its form, but is rather accidental for both *x* and *y*. Let us call them *thin* relations in Mulligan's sense.

Lowe has another, more metaphorical, explanation of this distinction between formal and thin internal relations: for the duration of the formal relation, the relata are "made for each other".[9] I would like to add a non-metaphorical distinguishing mark: thin relations seem to be in a proper sense derived from accidental aspects of their founding instances. *X*'s being bigger than *y* is derived from *x*'s (accidentally) being *F* and *y*'s (accidentally) being G. For formal internal relations, by contrast, we cannot allow this accidental one-way-derivation. Since formal relations concern the nature or the form of their relata, they are by definition ontologically indispensable to those relata. This is self-evidently the case with characterization. The characterization-function of a mode *F* is not derived from an "accidental" aspect of *F*, but rather is fundamental to it. Formal relations, to draw once again on Jonathan Lowe, are not derivative, but "too fundamental ... to be something in the world – an element of being – because it is that without [which] there could be no beings and so no world."[10]

Let me add another important point that holds for formal, but not for thin, relations: Formal relations have founded some sort of ontological dependence between their relata. Mulligan and Lowe consider this aspect too. Mulligan speaks of an "involvement" which concerns all formal relations,[11] and Lowe describes

8 Lowe (2006), p. 48.
9 Ibid., p. 47.
10 Ibid., p. 49.
11 Cf. Mulligan (1998), p. 345.

the co-occurrence in question as something to be apprehended from the other side, that is, from dependence: "... all dependence relations are, in a certain sense, *founded* upon ... formal relations – relations which are, for this reason, ontologically more basic than the dependence relations themselves."[12]

Space does not permit taking sides between these two approaches. Rather, we are highlighting the constitution-relation, and want to ask how defining constitution as such an internal, formal, dependence-founding tie may lend support to our claim that temporal relations are epiphenomena.

3.2 Constitution as formal relation

Before giving an account of constitution as a formal relation in the sense introduced above, let us avoid misunderstanding by a brief sojourn along the *via negativa*, saying what constitution is not. To put it roughly: we are not using "constitution" in the meaning that, for instance, is given by Lynne Rudder-Baker. According to her: "[t]he fundamental idea of constitution is this: when a thing of one primary kind is in certain circumstances, a thing of another primary kind – a new thing, with new causal powers, comes to exist."[13] The basic idea of constitution, according to Rudder-Baker, is that constitution is a relation that occurs between things which emanate on different levels of reality, where "thing" is understood as a technical ontological term comparable to the sense introduced above.[14] According to this sense of constitution, a thing of a primary kind *F* becomes a thing of a primary kind *G* (which is different of *F*). The constituted thing (of kind *G*), *y*, cannot be the same as the constituting one (of kind *F*), *x*. Emanation brings about something new.

Rudder-Baker uses the example of the relation between a statue and its stuff or its material. According to Rudder-Baker, a lump of bronze constitutes, under some suitable circumstances, the statue – that is, it lets the statue emanate from the bronze. But the constitution-relation does not occur only between lumps of matter and macro-things; it also occurs at other levels of reality, from the very "top" level to the very "bottom" levels of micro-physics, such as parcels of mass, atoms, quarks, etc.

The view developed in this paper is constructive, not polemical, and for this reason we will not object extensively with Rudder-Baker's concept of constitution.

12 Lowe (2006), p. 34.
13 Rudder-Baker (2007), p. 32.
14 Cf. Rudder-Baker (2007), pp. 33ff.

I just want to mention in passing that the multi-layer-picture of the world implied by Rudder-Baker's constitution-theory lays itself open to a number of objections.[15] One is that, according to my ontological scheme, lumps of bronze or other materials cannot be things in the sense introduced above, but rather *quasi-individuals* with indeterminate identity.[16] It is even less plausible to regard micro-phenomena like quarks and atoms as things – but that is another story.

To return to our theme of the theory of constitution: how can we understand it in a non-Rudder-Bakerian way with the tools I have already introduced? Three points can be made, the first two of which are implicitly non-Rudder-Bakerian. The third will mark the difference between my approach and Rudder-Baker's explicitly.

First, constitution is an *internal* relation and not an entity. The standard argument here is as follows: if constitution were an entity, what would relate it with the constituting and the constituted? Another, second-level relation? Should we regard these second-level relations as further entities? It is obvious that no matter how far back we go, the regress will extend. On the other hand, should we assume the second-level relations to be non-entities? This would stop the regress, but how could we argue against the entity-status of the second-level relation, if we accepted it for the first-level relation?

Constitution is internal, but, second, it is not a thin relation, but rather a *formal* one. Whatever constitution may be, it pertains to the nature of the constituted to be constituted. It is not accidental to it. The same holds for the constituting. Nothing can constitute something due to an accidental aspect.

Our third point is that the logical-formal character of constitution can at best be understood with reference to the *dependence that is typically founded upon* formal relations. This constitution-specific dependence relation has three characteristics. (i) It is *irreflexive*: nothing can constitute itself and thus create an autodependent founded entity. With regard to the problems with Rudder-Baker's concept of constitution, we can take this irreflexivity not only as an individual but also as a generic affair: no *x* of a kind or category *F* can constitute another *y* of the same *F*. (ii) It is *asymmetric*: if *y* is constituted by *x*, *x* cannot be constituted by *y*. No two entities can mutually stand in constitution-founded dependence. Here, too, this condition has a generic dimension: if an *x* of kind *F* constitutes a *y* of kind *G*, it is impossible that another *G* constitute *x* or another instance of *F*. Finally (iii), it is *non-transitive*. This distinguishes our concept of constitution definitively from

15 I think John Heil's arguments against a "multi-layer-picture" particularly convincing; see, Heil (2003), pp. 28f.

16 In the sense of what Lowe calls "quasi-objects", see e.g. Lowe (1998), p. 58.

Rudder-Baker's. If an x (of kind F) constitutes y (of kind G, not identical with F), neither y (nor another G) can constitute some z, which would then also be constituted by x. The dependence of the constituted y on the constituting x deprives y of being the constituent, and thus of being the dependence-basis, of some other z. Incidentally, this blocks any possibility of "multi-layer-models" of reality. For this reason, all that is constituted must be understood as something ontologically secondary, as an epiphenomenon. Any obscurity in this conclusion will be made explicit in the next section.

3.3 Events and epiphenomenal temporal relations

We now have the theoretical elements in hand to explore the way in which time and temporal relations are constituted by events.

Let me start by characterizing the constituting tie between events and temporal relations as internal. It follows that the constitution of temporal relations by events (where events are understood as unified sequences of continuously occurring parts or phases), is not an entity unto itself. There is no dyadic entity that exists between a sequence of event-phases and the temporal relations they constitute. It is the occurrence of sequences (or phases) that constitutes e.g. being earlier or being later. This is how the assumed immediacy between events and time, according to which events are paradigmatic temporal entities, are to be explicated. This explication precludes understanding the constitution of time by events in terms of adding something to reality.

The constitution of temporal relations by event-phase-sequences is formal internal, as opposed to thin internal, in the sense introduced above. It is due to the *nature* or the *form* of events, which are phase-sequences, that they constitute temporal relations. And it is in the "nature" of temporal relations to be constituted by events.

If the constitution of temporal relations by events is formal internal, then it is also – in accordance with the foregoing – a *dependence*-founding relational tie. That temporal relations are constituted by phase-sequences yields the result that temporal relations are ontologically dependent on phase-sequences. With our formal tools we can describe this kind of dependence so as to legitimate an anti-realistic or epiphenomenal account of the constituted.

First, we can state that the dependence between temporal relations and event-phases is *irreflexive*, since reflexivity can, for every constitutional tie, be excluded at both the individual and the generic levels. An event can be tied neither to itself nor to another event in virtue of the way which it is tied to any temporal relation. And nor can a temporal relation, in virtue of how it is connected with an event, be tied to itself or to other temporal relations. These observations are intuitively plausible. Moreover, their ontological implications are relatively unsurprising. Just to mention one such implication: if the suggested sort of irreflexivity applies, then no time can be an event, and no event a time. A realist about time must establish time as a category of being unto itself. But the strategy of construing time as a sub-kind of event-like entities, as we have seen, is blocked by our construal of constitution – which, although it does not *preclude* a realistic account of time, makes it less plausible.

The second formal aspect of constitution is individual and generic *asymmetry*. When events constitute temporal relations, the resulting dependence of temporal relations on events precludes this dependence from occurring between events and temporal relations. Moreover, the asymmetry of dependence brings us closer to an anti-realistic account of temporal relations, and consequently to an epiphenomenal interpretation of time. If events do not depend on time in an ontologically significant way, then what else does? If nothing depends ontologically on time, then what could render the very idea of being an entity, or the existence of temporal relations, intelligible? For it is downright unintelligible to deny that entities ground ontological dependence in at least some way.

This consequence is also supported by the third formal characteristic of constitution. It is *non-transitive*. If temporal relations are constituted by events, then such relations cannot constitute anything else that would have been regarded as the product of the constitution of events. Being constituted by events renders it impossible for temporal relations to serve as the constitutional basis of some other relata.

The conclusion of the foregoing is that, since constitution is an internal relation, there is no entity "between" events and time. Constitution is a formal internal. The reason is that events consist, essentially or "in their nature", of sequences of phases, which *constitute* time. This formal tie between events and temporal relations brings the latter into a kind of dependence on the former; this makes it impossible for temporal relations, themselves, to be the basis of constitution.

It is especially important to note that, since time cannot serve to ground any other form of ontological dependence, the alternative arises of considering time to be a sort of epiphenomenon. Standardly, because epiphenomena are phenomenally and, to some extent, theoretically indispensable, they are considered ineliminable even though they play no role in the causal nexus of the world. Some authors regard mental phenomena like qualia as epiphenomena. I do not intend to contribute to this debate. I shall merely suggest – with reference to time – that we regard time as indispensable to any theory of the world, but that we shift our focus from the causal to the *constitutional* inefficacy of epiphenomena. Because time and temporal relations are epiphenomena, they are indispensable products of constitution, but they cannot be the basis of some further constitution, nor can they ground any other kind of ontological dependence.

If epiphenomena are inefficacious, then, in spite of their (phenomenal) indispensability, they cannot be considered to belong to the basic structure of reality. Epiphenomena are not entities. This result fits well with the standard analytic understanding of epiphenomena. In our case, this would imply a kind of moderate anti-realism about time and temporal relations.

Constitution as a formal relation can be re-interpreted from this standpoint as well: Constitution is that formal internal tie which grounds a strong enough ontological dependence so that the dependent thing loses the status of entity and must thus be regarded as epiphenomena. In short: constitution ties entities together with epiphenomena. Events are entities which – being unified sequences of different phases – constitute epiphenomenal temporal relations.

It would be interesting to consider additional arguments for this claim, especially from the perspective of a thing- or substance-ontology; I would claim that such an ontology can only be logically consistent if it maintains a (moderate) non-realistic or epiphenomenal account of temporal relations. But this would lead us too far afield. In this paper, suffice it to consider one final issue: How can we apply this ontological position on time to infinity, God, and God's characteristic infinity?

4 The outlook

Classically, we think of "infinity" in two different ways: either as *temporal* infinity, or as non-temporal or *timeless* infinity.

If we consider time to be an epiphenomenon constituted by events, temporal infinity can be construed as implying an infinite or endless chain of events, just as the ancients thought that circular movements (of planets) would be infinite. Some modern cosmologists take up this ancient intuition too, applying it in the theory of an endless cosmic loop. But the validity of such a model is outside the scope of this paper. If we want to include temporal infinity in the way we understand the world, then we need this sort of endless loop consisting of endless chains of events. If we are skeptical about such loops, then we must also be skeptical about the temporal infinity. I leave aside the question of the contrary: finite timeless universes, i.e., universes in which, accidentally, no events occur; instead I will focus on another classical concept of infinity: non-temporal, timeless, or eternal infinity. This is the notion we will need to characterize divine infinity.

If time is an epiphenomenon based on events, then a subject which is essentially not involved in any event would in general be non-temporal (alternative expressions include "timelessly infinite" or "eternal"). If a subject were a *perfect being* (otherwise known as an *actus purus* or an absolute *simple being*[17]), then that subject could not possibly be involved in any event. The reason is that, since all events are either beginnings, endings, or changes, involvement in one of them would by definition preclude the subject's perfection or its ability to be simple or an actus purus or being simple. Granted that God is a perfect being, an actus purus, and simple, he must not be involved in any event, which would mean he would have to be timeless infinite or eternal.

It is worth adding that, if time is epiphenomenal, timelessness is no flaw. As an epiphenomenon, time would have no being. Just as the lack of a privation would be no deficiency in being, neither is the lack of something that has no being in and of itself.

God's being can be regarded as timelessly infinite, i.e., eternal. In addition to this claim, our ontological theory of time yields other implications for thinking about God. In particular, we can fruitfully use it to discuss the puzzle of God's relations to his creation, which indeed is temporal. Examples of such relations include his *knowing* temporal things and (especially) his *acting* in a temporal cosmos.

In the following I will draw (liberally) on the interpretation put forward in William Hasker's *God, Time, and Knowledge* (especially chapter 8), where Hasker argues for the principal intelligibility of God's timeless knowing and acting in a temporal world.[18] I would like to add my ontological interpretations to Hasker's argument. The least problematic application seems to be God's knowing, since knowing is "not a time-consuming activity" (ibid., p. 152). As I would phrase it, knowing is neither necessarily nor definitionally an event-dependent occurrence. Not every knowing is the result of a process in time; indeed, such a result would presuppose a kind of "learning-process" which, ontologically speaking, amounts to an imperfection. On the contrary: "this is a limitation of our finitude, which obviously does not apply to God" (ibid.). In other words: That a knower *x* knows *y* does not contradict the claim that *x* is timeless, even if *y* is tensed.

Acting is ontologically more significant but also more problematic. Let us suppose (with Hasker) that the paradigmatic divine action in our world is God's "preservation of the world in being from moment to moment" (ibid., p. 152). How

17 To the close conceptual connection between simplicity and immutability see Hasker (1989), pp. 182f.

18 That Hasker finally takes a critical view on God's timelessness, especially with regard to the problem of God's foreknowledge and free will (see ibid., chapter 9) can be neglected here.

then can we understand the preservation of temporal beings by a timeless preserver? Hasker argues that such a procedure is intelligible. To do that he first discusses the intelligibility of a non-spatial God who is able – from outside of space – to bring about effects in space; Hasker concludes: "Just as the non-spatial God can act outside of space so as to produce effects at every point in space, so the timeless God can act outside of time, that is, in eternity, so as to produce effects at every point in time." (Ibid, p. 154) The relevant distinction is between the act itself, which in the case of the timeless God is itself timeless, and the effects of the act, which may be temporal: "… the temporal characteristics of the effects of divine actions need not characterize the actions themselves. The act of preserving a temporally extended universe need not itself be temporally extended." (Ibid., p. 158) Our sketch of the epiphenomenality of time helps us explain why it is consistent to maintain that God timelessly preserves temporally extended entities (such as things and their modes, states, and events): That these effects of God's preservation are temporal does not affect God's act. The reason is that the temporality of things, modes, states, and events results from the ontological function of events, which is to constitute the epiphenomenon of time.[19] God preserves events, but he need not and indeed cannot preserve the epiphenomena that are constituted by these events. The same may hold for things and space: God preserves things, but not the epiphenomena that they constitute. This is no deficiency of God's, because epiphenomena are not entities – they have no being at all. There is no deficiency in not creating or preserving non-entities.

In summary, this paper aimed to develop a three-fold theory of temporality: (i) Events are the immediate constituents of temporal relations. This is what it means to say that they are four-dimensional. (ii) Temporal relations are interpreted in a moderately anti-realistic way, namely as epiphenomena. (iii) What links (i) and (ii) is the assumption that constitution is a formal internal relation (which is irreflexive, asymmetric, non-transitive); the further theoretical significance of this link, especially for a substance-ontology, cannot be pursued here. Finally, (i)-(iii) are compatible with (traditional) interpretations of infinity and of God's eternality.

19 How the "temporal shapes" of things, modes, and states depend on events and their function to constitute temporal relation, is another ontological question which must be left for another occasion.

Bibliography

Rudder-Baker, L. (2007), *The Metaphysics of Everyday Life*, Cambridge: Cambridge University Press.
Hasker, W. (1989), *God, Time, and Knowledge*, Ithaca/New York: Cornell University Press.
Heil, J. (2003), *From an Ontological Point of View*, Oxford: Clarendon Press.
Lewis, D. (1986), *On the Plurality of Worlds*, Oxford: Blackwell.
Lowe, E. J. (1998), *The Possibility of Metaphysics*, Oxford: Clarendon Press.
Lowe, E. J. (2006), *The Four-Categorial Ontology*, Oxford: Clarendon Press.
Mulligan, K. (1998), "Relations – Through Thick and Thin", in *Erkenntnis*: 48 (Extra-Volume: Analytical Ontology, edited by. Ch. Kanzian und E. Runggaldier), 325–353.
Szatkowski, M. (ed.) (2015), *God, Truth, and other Enigmas*, Philosophical Analysis 65. Berlin/Munich/Boston: De Gruyter, 183–196.

Tomasz Kąkol

In Defense of Presentism and an Extratemporal God

I begin with some definitions. "Presentism", defined here as the claim that a time flow objectively exists along with an objective difference between past, present and future, is the stance defended by a minority of contemporary ontologists of time. Another prominent view, eternalism, claims that both the difference between past, present and future and the flow of time are merely subjective. An extratemporal God (well-known in traditional rationalist metaphysics[1]) is often thought of as the opposite of "the living God": philosophers since Aristotle (not to mention common sense) have held that change implies time and therefore, extratemporality entails unchangeability.

One can argue that my definition of "presentism" is idiosyncratic; specifically, one usually understands this doctrine as stating that only the present (but neither the past nor the future) exists. This is true,[2] but I deliberately depart from this unfortunate custom since it creates an ontological straw man that is too easily refuted.[3]

In my paper I defend presentism using two positive and two negative arguments: 1) the presentist "now" is assumed in contemporary physics, arguments to the contrary notwithstanding; 2) presentism has explanatory power, and can explain many phenomena such as the permanent change in our temporal perspective and our concern about the future and the asymmetry of biological growth[4]; 3) the putative explanation of the intuition of time flow is ungrounded, and the same applies to both the entropic theory of time and the causal theory of time; and 4) presentism neither implies relative existence (when combined with special relativity) nor is subject to the notorious "how fast does time flow" objection and the alleged rejection of the so-called "truthmaker principle".

1 One of the main reasons for holding that God is extratemporal is that a temporal God cannot both be omniscient and respect human freedom. If humans act freely, then the outcomes of their deeds cannot be predicted with 100% certainty, whereas an extratemporal being need not *predict* anything.

2 See, e.g., Sider (2001), Hansson Wahlberg (2009) or Grygianiec (2011)

3 See, e.g., Dummett (2004), p. 74 – if presentism (as "usually" understood!) is true, then the present "is a mere boundary between past and future. But a boundary can exist only if that which it bounds exists." So, nothing would exist!

4 Such as the change "from the kernel to the mature tree". See Gołosz (2011), p. 152.

https://doi.org/9783110594164-004

Although I owe those arguments to Jerzy Gołosz (the most vigorous proponent of presentism in contemporary Polish philosophy), the ontology of time that I propose is more Ingardenian in spirit and more moderate: in particular, I disagree with Gołosz's theses that (a) presentism entails endurantism[5] and (b) the famous solutions of Einstein's field equations of gravitation with closed time-like curves do not entail that presentism only contingently applies to our world.[6]

As regards God, we have a mathematical model of tenseless dynamics (Michał Heller) that not only answers the objection that I mentioned but also makes sense of the traditional "dark" conception of *creatio continua*.

*

Ad 1. A. Einstein famously wrote to the widow of his friend that "the [objective] distinction between past, present and future is only an illusion, even if a stubborn one"[7]. But this remark raises a question: why do physicists stubbornly talk – as, for example, Stephen Hawking does in his *A Brief History of Time* – about the *present* rate of the expansion of the universe or the *present* value of the density of matter[8] or the *present* value of background radiation[9] or the *present* bond between electric and magnetic fields[10] and so on. Einstein is suggesting to the widow that eternalism is true, but if it is, why do physicists often talk as if presentism is true. In short, presentist language is ubiquitous in contemporary physics.

Ad 2. As time flows, we undergo a permanent change in our temporal perspective; some (Einstein, Hawking in the book mentioned[11] and, interestingly, even poets such as R. M. Rilke in his *Letters to a Young Poet*[12]) try to explain this by claiming that moving bodies such as the Earth in fact move along a straight path in four-dimensional spacetime. However, this is inconsistent with the four-dimensional, eternalist picture of a static or block universe. To put it another way, if we talk about moving in a n-dimensional structure, this raises the objection that we thereby are just assuming an objective time.

Moreover, eternalists need to explain why we have more interest in the future than the past if eternalism is true. One attempt, by Paul Horwich, grounds this

5 Conf. idem, p. 9.
6 Conf. idem, p. 133.
7 The quote is from Davies (1995), pp. 70, 76.
8 See, Hawking (1998), p. 27.
9 See idem, p. 98.
10 See idem, p. 98.
11 See idem, pp. 20 and 68.
12 Conf. Rilke (1977), p. 70.

concern in biological fitness, but the truth is that this is so *only* because there is asymmetry between the past and the future (the past is unchangeable), which calls for further explanation.[13]

Ad 3. The usual explanation of the alleged illusion of time flow is that our cognitive processes purportedly produce this illusion. But the problem is that they are, simply as *experienced processes*, temporal and dynamic entities.[14]

As for probably the most popular theory of time, namely the entropic theory of time, if time is identified with or at least correlated with (or determined by) the growth of entropy, then what is happening in regions in which reversible processes hold since the statistical character of the second law of thermodynamics allows for such realities? Is time static there? Or does the flow of time also reverse? The former can reasonably be rejected. Some people say that the problem is that, viewed in this manner, the second law of thermodynamics becomes trivial.[15] I think otherwise, since the fact that water is H_2O is not trivial. A much more serious difficulty is why we remember the past and not the future if we do not concede, following Hawking, that if entropy were decreasing, then we would remember the future![16] Indeed, his firm conviction unfortunately looks like fantasy.

The causal theory of time, another prominent view, also faces objections. First, as was observed long ago, causality at most presupposes time, but it is not implied by it[17] (at most, since we should not exclude a priori non-temporal causality). Second, if causality is guaranteed by the laws of physics, which concern (nowadays four distinct) fundamental forces (strong nuclear, weak nuclear, gravitational and electromagnetic), these forces – apart from the second one which can be ignored from the macroscopic point of view – are insensitive to the direction of the passage of time, so to speak.[18]

---

13 See Gołosz (2011), p. 31.

14 See idem, pp. 24, 26, 32. Some readers may accuse me of contradiction since I mentioned tenseless dynamics and later I will talk about the "Gödelian world" which is in motion but without time flow. I reply that the model of tenseless dynamics comes from microphysics and should not be extrapolated to the world of such "medium-sized phenomena" as our mental states and processes (it is consistent with the utility of this model in showing that an extratemporal God *can* be "dynamic" since we do not have any experiential access to Him/Her, whereas we do have access, at least partly, to our mental states – through introspection). Similar points should be made about "Gödelian world" as we will see in a while.

15 See Gołosz (2011), pp. 162f.

16 See Hawking (1998), pp. 73 and 75.

17 See Gołosz (2011), pp. 162, 171.

18 See idem, p. 176.

Ad 4. Presentism does not entail relative existence when combined with special relativity; presentness is just relativized to the frame of reference (recall that the "relativity" in special relativity does not imply non-objectivity). As for the objection that presentism contradicts special relativity because different objects in different frames of reference can be regarded as simultaneous or co-present or coexisting (in the tense sense of the term)[19] – it should not be a reason for worry since the causal structure of spacetime is saved across the frames of reference.[20] Similarly, the alleged problem of the truthmaker is straightforwardly solved: the truthmaker of, say, starting World War II *does not occur* (or, to stress the tense sense of the term, *is not occurring* now), since it has already ocurred.[21]

To answer the question "how fast does time flow?" we need to introduce more ontology. Following E. Husserl, R. Ingarden makes a distinction between *pure* and *empirical* possibilities (and other modalities), restricting ontological considerations to the former only. Closing the first volume of his monumental "Controversy over the Existence of the World" (I mean here the last, complete edition, i.e. from 1987), Ingarden gives the characteristics of eight possible modes of real existence (as opposed to the absolute, the ideal and the purely intentional). Ingarden divides these eight modes into three groups called the *presence*, the *past* and the *future*. In other words, time in general (in specie) is for Ingarden the mode of existence of real objects (or, equivalently, the mode of real existence or the real mode of existence – according to Ingarden, all these expressions have the same meaning, although he prefers the phrase "the real mode of existence"). Remembering that Ingarden states that every object is a triunity of matter, form and the mode of existence (matter and form understood in a non-Aristotelian way), he follows the tradition according to which there is a very close link (to say the least) between the concept of time and the concept of existence. To highlight this connection (as it is found in the ambiguity of the English word "presence" and in many other languages – but, interestingly, not for example in Polish) I propose to name – in a Heideggerian manner – the past *wasence* and the future *willbence*. Next, Ingarden divides real beings into three classes: objects enduring through time (substances, using traditional terms, or endureres), processes, and events. They have different forms and, consequently, different modes of existence: enduring through time or short endurance, becoming and (unique) occurring or occurrence respectively (unique, since Ingarden is convinced that events never recur or return). In effect, we have the following table of time:

---

19 See Hansson Wahlberg (2009), pp. 21–23.

20 See Gołosz (2011), p. 127.

21 Conf. idem, p. 98.

time	presence:		
	present endurance	present becoming	present occurrence
	past (wasence):		
	past endurance	past becoming	past occurrence
	future (willbence):		
	future endurance	future becoming	future occurrence
	↓ of endurers	↓ of processes	↓ of events

In this way – contrary to, e.g., St Augustine – we can clearly tell what time is: first, time is always the time *of* something. For example, the past of a pen is its past endurance, the presence of this process of my writing is its present becoming and the future of my departure to Budapest (assuming that it is an event – some people could argue that it is more of a process) is its future occurrence. Secondly, the Ingardenian conception can meet the challenge of the notorious "how fast does time flow" charge. To begin with, the becoming of the process of reddening, say, of a spoon for babies (there are such spoons, made of material that change color depending on the temperature of milk or another kind of drink) can be faster or slower, but when we choose the model time – for example, one becoming of a full revolution of the pointer of a certain gadget/machine – we can *measure* time saying for example that the reddening process of the first spoon is twice as fast as the reddening process of the second spoon, since the former takes up half a model time, whereas the latter takes up the whole of it. If a critic were to say that we have now introduced a vicious circle or a regress, as we can ask about the velocity of the becoming of this full revolution, we could reply that if she were right we could not ask for example "how long is the length of the model?" (the late Wittgenstein thought so but he was rightly criticized by Kripke). The only inconvenience is linguistic, since it sounds bizarre that the mode of existence can be faster or slower.

*

Data from physics suggest that every presentist ontology of time, Ingarden's included, should be limited for at least two reasons. First, in 1924 Cornelius Lanczos, later Einstein's assistant, discovered some very peculiar solutions of Einstein's field equations of gravitation. They were in a sense rediscovered by Willem Jacob von Stockum in 1937, and then by Kurt Gödel in 1949, and, due to the latter's famous name, they are often referred to as "Gödelian solutions"[22]. In general relativity we treat space and time inseparably, as *spacetime*. In this way we can represent within this structure *histories* of objects with non-zero rest masses (such as atoms or electrons) as certain *curves* (physicists call them "timelike curves"). Imagine a caricature of a Heraclitan (or Stoic or Nietzschean) world of the so-called eternal return that would consist of two spatial dimensions only *and* two small point masses rotating around each other. When we add a temporal dimension, we obtain two timelike curves similar to a double helix (very well-known from biology). However, the peculiarity of Gödelian solutions is that there are closed timelike curves there that entail the *prima facie* possibility of time travel. It should be stressed here that Gödel's universe cannot model ours, since, apart from its global rotation, it is static, whereas ours is expanding (moreover, Gödel's world is simplified, as its spacetime is homogeneously occupied by dust with constant density); nevertheless, it shows that presentism is probably contingently only true of our world. The reason is simple: closed time excludes both the time flow *and* the division into past, present and future. Imagine that we choose a point on a closed timelike curve as "now". According to presentism, what was before, there is no more, but before was the very same point, which amounts to a contradiction. As for time travel, it does not make sense to interpret it as going back to the past, since this is not "the past" in the presentist sense. When someone thinks that there is a trick here in that in this way we have only shown that the Gödelian universe is inconsistent, recall that something similar – and equally paradoxical – holds in special relativity concerning photons (that is, quanta of electromagnetic force): according to special relativity, there is no lapse of time in them. Nevertheless, photons are constantly in movement, indeed the fastest physical movement! Consequently, presentism in all probability is *both* contingent and local. Its locality is additionally suggested by several cosmological models of (semi)quantum gravity. According to one of them, that is, the well-known (and criticized) model of Jim Hartle and Stephen Hawking from 1983, there is no time when we cross the famous Planck's thresholds (Planck's time $\approx 10^{-44}$s, Planck's length $\approx 10^{-33}$cm, Planck's density $\approx 10^{93}$g/cm^3). The much more interesting one is the more radical: there is neither time nor space when we cross these magic boundaries; this view has been popularized in Michael Heller's book under the very meaningful title *The Beginning is Everywhere* ("and at anytime" should be added) in 2002 (in Polish).[23] This model not only makes "desingularization of space-time" successful (as Hawking's

22 See Heller (2012), pp. 48–50.
23 See Heller (2002).

does)[24] but also solves the so-called horizon problem without ad hoc inflation (almost every contemporary cosmological model assumes mysterious inflation – that is, the rapid expansion of the universe) and, in addition, explains the notorious Einstein-Podolsky-Rosen paradox of non-locality.[25] Setting aside mathematical details, Heller's model uses non-commutative, pointless space. Since ("standard") space and time imply locality, because they consist of points (interpreted as moments when time is concerned) or extended simples, as some – Ingarden included – think, which are local entities, non-locality *excludes* both space and time.

The consequences of this approach, if Heller is right, are philosophically far reaching: it not only denies naïve materialism stating, as David Armstrong would put it, that the totality of being is no more than the *spatiotemporal* system, and denies the phenomenologically and commonsensically obvious proposition that dynamics or change implies time, but also entails that Ingarden's conviction that time is the mode of existence of all real objects is unsupported. The consistency of tenseless dynamics is obtained by defining in the space that Heller uses analogs of (force) vector fields[26], whereas the similarity to the traditional doctrine of *creatio continua* is obvious: when we arbitrarily choose at time *t* any physical thing *A* in our vicinity (say, the table I am sitting at) and we are mentally (i.e., using our imagination) approaching Planck's length, we are approaching Planck's density and the region (or "the region"), where there is neither space nor time – and this is just the region sought by cosmologists debating about the so-called Big Bang and its vicinity. Thus, the "true" beginning, or the ultimate origin of my table, is borne out at *t*.

Of course, this model also poses questions traditionally directed to the theory of *creatio continua*, such as whether we have in effect overdetermination (since prima facie my table would be both caused by the carpenter somewhere in the past, and yet constantly borne out), but this is a topic for another article.

Bibliography

Davies, P. (1995), *About Time: Einstein's Unfinished Revolution*, London: Penguin Books.
Dummett, M. (2004), *Truth and the Past*, New York: Columbia University Press.

24 The so-called singularity is known even among laymen just as black holes are.

25 Concerning those problems see Heller, idem, chs. 10f. Notice especially that the horizon problem is interpreted as also indicating non-locality.

26 See idem, ch. 9.

Gołosz, J. (2011), *Upływ czasu i ontologia* [*The Flow of Time and Ontology*], Kraków: Wydawnictwo Uniwersytetu Jagielońskiego.

Grygianiec, M. (2011), “Trwanie w czasie” [“Persistence in Time”], in *Przewodnik po metafizyce* [*A Guide to Metaphysics*], edited by S. T. Kołodziejczyk, Kraków: WAM, 211–276.

Hawking, S. (1998), *A Brief History of Time*, New York/London/Toronto/Sydney/Auckland: Bantam Books. Text available at: https://www.pdf-archive.com/2014/07/24/a-brief-history-of-time/ (access: 1.09.2016).

Rilke, R. M. (1977), *Lettres à un jeune poète*, Paris.

Hansson Wahlberg, T. (2009), *Objects in Time. Studies of Persistence in B-Time*, Lund University (doctoral dissertation).

Heller, M. (2012), *Ostateczne wyjaśnienia wszechświata* [*The Ultimate Explanations of the Universe*], Krakow: Universitas.

Heller, M. (2002), *Początek jest wszędzie* [*The Beginning is Everywhere*], Warszawa: Prószyński i S-ka. In English this model was presented in the article “The Emergence of Time”, in Physics Letters A, 250(1-3), 48–54.

Sider, T. (2001), *Four-dimensionalism. An Ontology of Persistence and Time*, Oxford: Clarendon Press.

Srećko Kovač

Concepts, Space-and-Time, Metaphysics (Kant and the dialogue of John 4)

1 Introduction

In an empirically oriented common-sense ontology, first-order concepts are expected to be "concrete" and to denote sensible objects given in space and time, while other, "abstract", concepts should denote words, sentences, sets, numbers, or concepts themselves, possibly of a questionable ontological status, or, moreover, conceived merely as a manner of speaking, subjective representations, or "ideas" without an actually corresponding reality. In a formalized presentation, such an empirical theory would have a model comprising a first-order domain of sensible objects denoted (possibly in n-tuples) by predicates. The domain itself and the relations on the domain, as well as syntactic "objects" (terms, predicates, formulas – replacing concepts and judgments), remain abstract, metatheoretical entities that are not empirically given for the object theory. Besides, the domain and the relations on the domain may appear as members of a second-order domain if the formalization is extended to a higher-order setting, but, of course, this still does not make the first-order domain and the relations on it themselves empirically existing objects.

Some essential features of abstract, model-theoretic, concepts of a possible empirical theory are traceable back to Immanuel Kant's "transcendental logic" (with some characteristic differences).[1] Against this background, we examine the objective reality of the abstract concepts involved, putting them in the context of a possible religious experience as presented in the text of John 4.

2 From metatheory to metaphysical theory

It can be recognized that Kant's theory of transcendental ideas serves as a sort of first-order model for empirical reasoning and knowledge, where transcendental ideas represent three sorts of totalities of conditions of empirical knowledge:

1 For some significant connections of Kant's logical theory with modern logic, see, e.g., Achourioti and van Lambalgen (2011) and Tiles (2004).

https://doi.org/9783110594164-005

(a) the totality with respect to a subject ("complete subject",[2] never occurring as a predicate, B 379): "I" ("mere consciousness", "determining Self"), which thinks, is the metatheoretical subject "X" of all thoughts (e.g., of concepts), which are its predicates (B 404, A 402);[3]
(b) the totality of the "series" of conditions ("world") of an empirically given object: each such object is possible only if the whole series of its conditions, too, is in some way already given (B 436);[4]
(c) the totality of concepts as predicates ("the sum total of all predicates") – as if comprised in some common "ground" (B 607): "the most real being" (*ens realissimum*).[5]

These "transcendental" ideas do not belong to empirical knowledge as an object-theory, but to its metatheory. Kant further specifies this by assigning those ideas a non-constitutive, regulative (and heuristic, B 644) role for empirical knowledge.[6]

We will now focus on some structural similarities between Kant's system of transcendental ideas and the conceptual structure of religious knowledge (religious belief)[7] as presented in Jesus' dialogue with a Samaritan woman in John 4.[8] From the standpoint of religious knowledge, transcendental (metalogical) concepts obtain their specific objective reality and become metaphysical concepts

2 Kant (1910–), Vol. IV, p. 330.

3 "I" is not a concept (or any representation) of an object, but just a general "form" of the knowledge of an object, since, as Kant points out, only by means of it do "I think anything" (B 404; cf. "I" is "that which I must presuppose in order to cognize any object", A 402). In this sense, we find "I" replaced in a standard first-order model simply by a chosen set of objects (domain). According to Kant's theory, the application of "I think" is restricted to the "manifold" (*Mannigfaltigkeit*) of what is given in a sensible spatio-temporal intuition. Let us note that the unity of a concept in Kant's (intensional) theory originates from the "analytical unity of consciousness", which "pertains to" the concept, while the unity of a concept (predicate) in a standard (extensional) first-order theory model-theoretically derives from the set itself (a subset of the domain) that is assigned to the concept as its extension.

4 This is in accordance with the "principle of reason": "if the conditioned is given, the whole sum of conditions, and hence the absolutely unconditioned, is also given, through which alone the conditioned was possible)" (B 436). In standard first-order model theory, element (b) of Kant's model is replaced by a relational structure that is imposed on the domain by the interpretation of relation symbols and complemented by the conditions of the satisfaction of formulas.

5 Cf. Kant's "principle of thoroughgoing determination", according to which each object should be determined with respect to each concept (B 599–600). In a first-order theory, the interpretation of one-place relation symbols replaces element (c).

6 For Kant, transcendental ideas are "regulative principles for the systematic unity of the manifold of empirical cognition in general" (B 699).

7 See on religious belief and knowledge in the introductory chapter of Kovač (2015a).

8 For a theological interpretations of this episode, see, for instance, Jojko (2012), Botha (1991) and Varghese (2009); for a theological-historical context, see, e.g., Novakovic (2013). Here, we further elaborate our initial analysis in Kovač (2011).

(soul, world, God); in addition, the application of concepts in general extends to the realm of non-sensible objects ("noumena").[9]

We first summarize the progress of Jesus' dialogue with the Samaritan woman (with a slight rephrasing) in Figures 1 and 2.

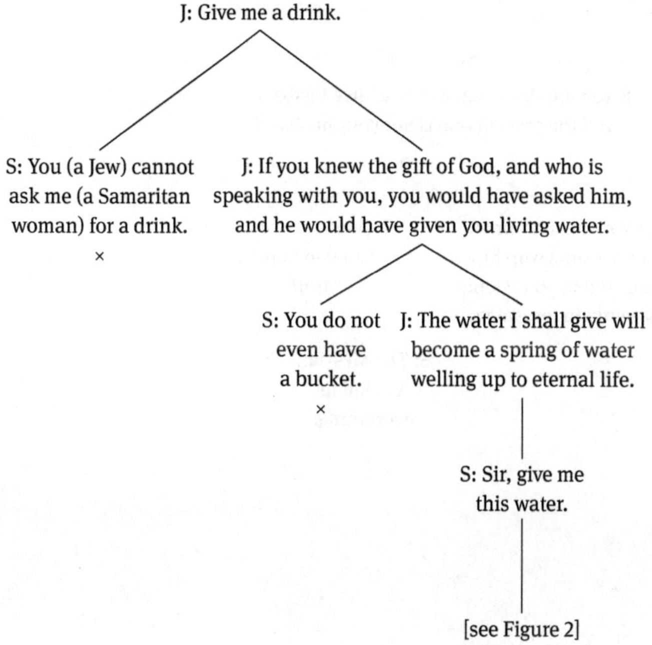

Fig. 1

We now briefly informally analyze the dialogue in John 4, comparing it with Kant's system of transcendental (metatheoretical) ideas.

9 Cf. B 395, footnote, with Kant's following critical remark: "Metaphysics has as the proper end of its investigation only three ideas: *God, freedom, and immortality* [...] The insight into these ideas would make *theology*, *morals*, and, through their combination, *religion*, thus the highest ends of our existence, dependent solely on the faculty of speculative reason and on nothing else". According to Kant, in the process of work we should proceed "from what experience makes immediately available to us, from the *doctrine of the soul*, to the *doctrine of the world*, and from there all the way to the cognition of *God*".

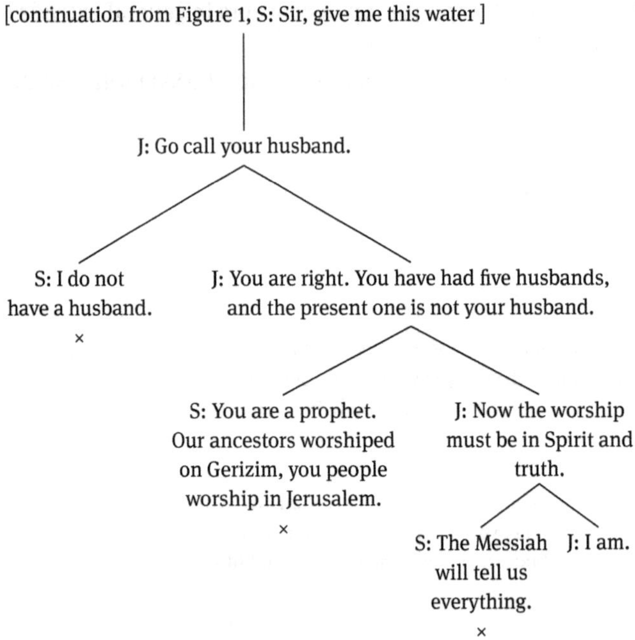

Fig. 2

2.1 Self

The idea of "self" ("I", "you", implicit in "we"), with knowledge, belief, speaking, and being as belonging to it, explicitly occurs and has an essential role in the dialogue of John 4. Two agents (selves), Jesus and a Samaritan woman, are engaged in the dialogue that is advancing step by step through logical reasoning and through a gradual evolving of new knowledge.

The logical idea of "I think" is recognizable where "I" is mentioned as a subject of epistemic (mental and verbal) acts.[10] In particular, the logical aspect of "self" is recognizable in the fact that the dialogue proceeds through the considering of and finding solutions for the contradictions appearing during the conversation. As will be seen, contradictions, as a means of a possible questioning, or

10 Cf., for example, "if you knew the gift of God and who is saying to you [...]", John 4:10; "you are right in saying [...]", John 4:17; "what you have said is true", John 4:18; "I can see [theōrō] that you are a prophet", John 4:19; "we worship what we understand", John 4:22; "I know that the Messiah is coming", John 4:25.

at least of a clarification, of the theses that appear in the dialogue, are the main logical vehicles of the conversation between Jesus and the Samaritan woman (see the next subsection).

In the Gospel, "self" is not just an abstract, regulative idea, but denotes a being, at first only an empirical being (a Jew, a Samaritan woman; John 4:9),[11] and eventually a being "in Spirit and truth".[12] According to Kant, in metaphysical psychology a paralogism occurs that is based on the non-justified assumption of an intuitive givenness of some persistent "self", which leads to the "inference" that "self" is a substance (B 411).[13] In distinction, the Gospel proposes a justification of the existence of "self" as a permanently ("eternally") given subject of religious (non-sensible, spiritual) experience.[14]

---

11 Cf. Kant's "empirical unity of consciousness": "One person combines the representation of a certain word with one thing, another with something else; and the unity of consciousness in that which is empirical is not, with regard to that which is given, necessarily and universally valid" (B 140).

12 It does not suffice for a worshiping agent to be in space and time, since the worship should take place "in Spirit and truth": "God is Spirit, and those who worship him must worship in Spirit and truth" (John 4:24). The spiritual *existence* of "self" in truth is explicit in Jesus' enunciation of his own being: "I am [the Messiah], the one who is speaking with you" (John 4:26; cf. subsection 2.3 below).

13 "Thus if that concept, by means of the term 'substance', is to indicate an object that can be given, [...] then it must be grounded on a persisting intuition as the indispensable condition [...] through which alone an object is given [...]. But now we have in inner intuition nothing at all that persists, for the 'I' is only the consciousness of my thinking." (B 412–413).

14 Even in Kant's moral philosophy, "self" is merely a postulate, not a concept that denotes objective reality. If we want to avoid the paralogism of metaphysical psychology (Kant, B 411) by means of a new sort of knowledge proposed in John 4, we get the following correct syllogism:

> What cannot be thought otherwise than as subject does not exist otherwise than as subject, and is therefore substance.
> A thinking being, considered as existing in Spirit and truth, cannot be thought otherwise than as subject.
> $\vdash$
> A thinking being, considered as existing in Spirit and truth, exists only as subject, i.e., as substance.

The expression "as existing in Spirit and truth" replaces Kant's "considered merely as such". In this way, "thinking" ("be thought") in the middle term is in the minor premise, too, understood with respect to a given (existing) object, not just with respect to the subject of thought (B 411, note).

2.2 World

We now proceed to a comparison of the dialogue of John 4 with Kant's transcendental idea of world – the totality of a series of conditions of empirically given objects and states (B 391) – and with Kant's corresponding cosmological antinomies. We will see that Kant's four conceptual aspects of a possible world totality – regarding (a) the composition of the whole of space-and-time, (b) the divisibility of matter, (c) causality, and (d) dependence in existence (B 438–443) – can also be found in John 4 (in a somewhat different way), as emerging one after another in the contradictions and their solutions unfolding during the dialogue.

Antinomies

Kant's antinomies arise from the question whether a given object and its state have some first condition (be it immediately given or not) or whether the series of its conditions is infinite (B 445–446). However, in John 4 it seems to be assumed that the world has the beginning in all four aspects (a) – (d) mentioned above (the beginning of time and space, the origin of matter, the first cause, and the unconditioned existence). So, the antinomies in John 4 arise from the problem whether the first condition (beginning) is immediately given (present) or whether it is given only through a (possibly long) series of intermediate conditions.

(a) Extension of space and time

Jesus: Give me a drink.
Samaritan woman: How can you, a Jew, ask me, a Samaritan woman, for a drink? (For Jews use nothing in common with Samaritans.)
⊢
Samaritan woman: [Contradiction].
(Cf. John 4:7–10).

Is a drink immediately available to Jesus, or should he look for it elsewhere and from someone else? To a significant extent, this can be interpreted in terms of Kant's first cosmological antinomy (B 454, 455), taking a drink as representing life (cf. "living water", John 4:10), and, in connection with this, as symbolizing time and space (flow and places, i.e., history, of life): *do I find the beginning of my life, i.e., of my time and space, immediately here and now* (*thesis*), or *should I return to the origins of my* (Jewish) *past, and go* out of this (Samarian) place back *to the land of my origin* (homeland, Galilee) (*antithesis*)?

Thesis 1: The beginning of an agent's *a* time and the beginning of *a*'s space are always immediately present.
Antithesis 2: The beginning of an agent's *a* time lies in the (past) moment of the beginning of *a*'s life, and the beginning of *a*'s space lies in the (distant) place of *a*'s origin.

Instead of Kant's opposition between the finite and the infinite regress in space and time, here we encounter the opposition between the immediate presence of the beginning of time and space, and its indirect givenness by means of intermediate segments of time and space.[15] Another interesting distinction appears regarding the conception of the beginning of space: in Kant, it is conceived as a possible end (outer limit) of space, while in John 4 it seems to be understood as the origin of space in relation to "self" (i.e., as home, homeland).

(b) Divisibility of matter

Jesus: If you knew the gift of God and x such that Jesus=x,
you would have asked x and x would have given you living water.
Samaritan woman: Sir, you do not even have a bucket and the cistern is deep.
⊢
Samaritan woman: [Contradiction, except that Jesus is greater than Jacob].
(Cf. John 4:10–11).

Can water be reached from the deep well and given without a bucket? In other words, to come closer to the terms of Kant's second cosmological antinomy (B 440, 443), *can matter (reality) be immediately given, without any partitioning (thesis)*, or *is matter given only piecemeal, in portions consisting of some elementary units* ("a bucket") (*antithesis*)?

Thesis 2: Matter can be immediately given to an agent *a* without any partition of matter.
Antithesis 2: Matter is given to an agent *a* only in portions consisting of units.

Again, we note a difference: in Kant's antinomy, there is the opposition between the finite and the infinite partitioning of matter, whereas in John 4 the opposition is between the immediate givenness of matter, and the givenness of matter only by means of its partition (possibly as a long series of portions of matter).[16] Besides,

---

15 The problem of mediation is already announced in the introduction to John 4: "[…] he [Jesus] left Judea and returned to Galilee. He had to pass through Samaria" (John 4:3–4).

16 Cf. for instance, "Sir, give me this water, so that I may not be thirsty or have to keep coming here to draw water" (John 4:15). That is, it is assumed that water is usually available only as a

we remark that Kant conceived the divisibility of matter by assuming that a thing to be divided (possibly into simple parts) is already given. In John 4, in distinction, the question regarding matter ("water") is whether it is available at all prior to its partition (thus, it is not atomism which is the central problem).

(c) Causation

Samaritan woman: Sir, give me this water [i.e., living water, which will become a spring of water welling up to eternal life].
Jesus: Go call your husband and come back.
Samaritan woman: I do not have a husband.
Jesus: You have had five husbands, and the one you have now is not your husband. What you have said is true.
⊢
Samaritan woman: [Contradiction].
(Cf. John 4:14–18).

Has the Samaritan woman an immediately present first ground of her own wish to get living water – thesis; or is her wish grounded on the whole causal series of states and events going back from the present to the past times out of her reach (five past husbands, with the present "non-husband") – antithesis? We can recognize an analogy with Kant's (third) antinomy of the causality of freedom and an endless series of the preceding causal events (B 441–442, 443), modified here into the antinomy between *the possibility that someone freely determines her (his) own wish/will* (*thesis*; cf. the Samaritan woman's wish as a possible free beginning of a new causal series), and *the determination by a (long) series of the preceding states* (*antithesis*; the Samaritan woman's determination by her past states).

Thesis 3: An agent a can freely begin a causal series by means of a's wish or will.
Antithesis 3: An agent a is determined by the preceding causal series of states.

It is interesting to note that in this antinomy the concept of causal beginning, both in John 4 and Kant, is related to free will/wish. However, in John 4 the (Samaritan woman's) wish is, at first, not directly related to an action, but to someone else's (Jesus') will ("Sir, give me this water"). We will come later (solutions below) to a related crucial difference from Kant's conception.

(long) series of portions of water drawn from the well.

(d) Dependence in existence

Samaritan woman: Sir, I can see that you are a prophet.
Our ancestors worshiped on this mountain [Gerizim], but
you people say that the place to worship is in Jerusalem.
⊢
[Contradiction].
(Cf. John 4:19–20).

Are the place and time of the presence of God (in worship) necessary in itself, i.e., independent of any further condition – thesis; or are *the place and time of worship contingent* (e.g., Mount Gerizim for the Samaritans, Jerusalem for the Jews), *i.e., dependent, for example, on long sequences of different traditions*[17] *– antithesis*? The thesis is not explicit in the text, but is implicitly contained in the antithesis, which is in itself contradictory (the traditions mentioned are mutually exclusive),[18] as well as in the solution (see below).

Thesis 4: God can be immediately present to an agent *a* independently of any contingent place and time.
Antithesis 4: God is present to an agent *a* only in dependence of some contingent place and time.

We can compare this antinomy of John 4 with Kant's antinomy between the thesis that there is an absolutely necessarily existing being (the world itself, or some being outside the world), and the antithesis, according to which all beings exist contingently (cf. B 442, 443). In the context of John 4, the existence of God is already presupposed in the solution of the third antinomy (see solution (c) below). Hence, the fourth antinomy specifically concerns the question of the existence of God; in addition, it concerns His *presence* – not just some necessary existence, possibly completely separated from the events in the world.

17 As concluded in Novakovic (2013), p. 215, "the main difference between the Jews and the Samaritans was not their ethnicity or religiosity but the location of their cultic center".

18 "There is no doubt that the building of the Gerizim temple [...] was met with disapproval by the Jews." As to the Samaritans, "Mount Gerizim [...] continued to be regarded as the only legitimate place of worship". (Novakovic (2013), pp. 212–213).

Solutions

Solutions of the antinomies in John 4 differ from Kantian solutions in that they solve each antinomy by *affirming the immediate presence in Spirit of the (unconditioned) totality (thesis)*, while *confining the antithesis (the totality accessible only through a possibly long series of conditions) merely to the sensible empirical world*. Only thesis is knowledge ("seeing") and about the true reality, while antithesis is ignorance.[19] In this way, a duplication of concepts arises (explicitly or implicitly): physical water and spiritual water, physical well and spiritual well (spring), physical giving/receiving and spiritual giving/receiving, physical drink and spiritual drink, physical thirst and spiritual thirst, time and space of the physical world and "time and space" of "Spirit and truth", physical husband and true husband, the truth of the physical world and spiritual truth, father (e.g., "our father Jacob", John 4:12) and Father, worship in a physical world and spiritual worship, physical self and spiritual self. By the term "physical", we intend to cover both the "natural" and the "historical" objects and states-of-affairs.[20]

As is well known, Kant gave a positive solution both for the thesis and for the antithesis only in the case of the third and the fourth antinomies: reducing them to subcontrarieties by distinguishing "sensible" from "intelligible" causality and dependence of existence.[21] On the other side, he resolved the first and the second antinomy by negating the thesis as well as the antithesis: reducing them to contraries because of the impossibility of the contradictory subject: the whole of time and space cannot be given in time and space, the whole partition of a material thing cannot be given in the thing, since this thing, which is, according to Kant, only our representation, is really divided only to the point up to which it is at a moment actually divided in our representation.[22] Regarding intelligible causality, let us recall that only the moral causality of freedom (connected with the third antinomy) received in Kant, in his practical philosophy, the status of objective reality.

19 For instance, "[...] we know that this is truly the savior of the world" (John 4:42). "You people worship what you do not understand [*ouk oidate*]; we worship what we understand [*oidamen*]" (John 4:22). "[...] no one can see the kingdom of God without being born from above" (John 3:3).
20 See Kovač (2015a) for the distinction of naturalistic and historicist conceptions of knowledge in John 3.
21 B 566, 587–589 and *Prolegomena*, Kant (1910–), Vol. 4, pp. 343–347, §53.
22 Cf. Kant's clarification in *Prolegomena* (Kant (1910–), Vol. 4, pp. 341–342, §52c), and in B 545–555.

(a) The first antinomy (extension of time and space). The solution confirms the thesis that the beginning of time and space is immediately present (Thesis 1) – as a "gift of God", while in the sensible world the beginning of time and space should be looked for (back) in the time and place of one's origin (Antithesis 1). In the Gospel, Jesus gives the solution by distinguishing between sensible empirical water and "living water", which is a "gift of God",[23] and which he presently possesses and could give to the Samaritan woman (cf. the first premise of the second antinomy above, John 4:10, and Figure 1). We note a specific way in which Thesis 1 itself is confirmed: living water is immediately available simply by asking for it in the right way (as for a gift of God).

Elaborating this a bit further, Thesis 1 leads to the equivalence relation (accessibility in time) on space points on the ground of the outer infimum (God), which is immediately related to a chosen inside point (Jesus, who comes from God; see John 4:25–26), and on the assumption of transitivity[24] and euclidity.[25] Of course, in a sensible world, there could be barriers which prevent accessibility in space and time.

This solution underlies the second antinomy (see above): whether matter (the content of space and time) can be given in space and time immediately, without any partition, or only piecemeal?

(b) The second antinomy (divisibility of matter). The solution affirms Thesis 2, according to which an indivisible origin of matter is immediately present, prior to any partition. This solution restricts the validity of Antithesis 2 to the sensible world, where we encounter only a (possibly long) series of portions of dividable matter, and it relates the validity of Thesis 2 to non-physical matter and "eternity".[26] In the words of the Gospel, what Jesus will give is not water in parts (in

---

23 According to *Genesis* 1:1–2, "in the beginning", before the creation of light, there was "a mighty wind sweeping over the waters" as a "pre-creation state". See the comment on this place in (New American Bible (2011), p. 10, note *).

24 Transitivity is indicated, for example, by the above-quoted sentence "If you knew the gift of God [...]" (antinomy (b), also Figure 1). Besides, see later in the text (assuming that "water" is, in a way, present in words): "Many of the Samaritans of that town began to believe in him because of the word of the woman" (John 4:39).

25 As an indication, we remark that the Samaritan woman, as well as other Samaritans in her town, heard Jesus' words, and then the Samaritans confirmed their belief to her: "[...] they said to the woman: 'We no longer believe because of your word; for we have heard for ourselves, and we know that this is truly the savior of the world' " (John 4:42).

26 The possibility and the presuppositions of something like "eternal life", as well as the questioning of the received views on it, are the subject of the third and, eventually, the fourth antinomies.

buckets), from some gradually accessible distance (from Jacob's deep well), but the spring of water and of eternal life:

> Everyone who drinks this water will be thirsty again; but whoever drinks the water I shall give will never thirst; [...] [it] will become in him a spring of water welling up to eternal life. (John 4:13–14).

This solution leads to a new (third) antinomy: to live in dependence on the physical world and its causal sequences seems to contradict the possibility to access the origin of eternal life. The problem should be answered by the next solution.

(c) The third antinomy (causation). The solution includes the affirmation of Thesis 3, about the causality of one's own wish (the possibility of a new causal beginning) through a (liberating) relationship to truth and God, as well as the affirmation of Antithesis 3, about an agent's dependence on a (long) causal sequence (e.g., tradition, personal history), as confined to the sensible empirical world. Thus, the Samaritan woman, living under the conditions of her past life (a sequence of husbands or "husbands") faces this fact (truth), and converts to the worship of God (in a way, she recognizes that truth comes from God).[27] The conversion to truth and the worship of God should make her free and able to break the preceding causal sequence of events, and to start a new sequence.[28] Such a new start does not exclude, but could perhaps rather require, having a true husband.[29] In a still wider perspective, the whole long causal sequence (history) consisting of "fathers" and "ancestors" (e.g., Jacob, Joseph) is now being replaced with the one present cause: the Father, as God is named by Jesus (John 4:21).[30]

---

27 Cf. later in the Gospel: "I tell you what I have seen in the Father's presence", "you are trying to kill me, a man who has told you the truth that I heard from God" (John 8:38,40).

28 See also John 8:32, "[...] the truth will set you free".

29 A relation between husband and wife could be closely described by Kant's category of "reciprocity" (*Wechselwirkung*). Kant gives the "reciprocity" ("community") of the parts of a body as an example (B 112). This could be compared with the following place from *Genesis*: "a man leaves his father and mother and clings to his wife, and the two become one body" (Genesis 2:24). We can also notice here that the community of "man and wife" is a sort of a new causal beginning, discontinuity with past. Let us mention that there are strong indications in John 4 that Jesus could be understood, in some true, spiritual sense, as a bridegroom (Jojko (2012)). However, according to *Genesis*, the community relationship between man and wife changed after the Fall to the rule of man over woman ("he [your husband] shall rule over you", Genesis 3:16); this one-sided causality could be related to the Samaritan woman's past husbands. The whole dialogue might be seen as a reversal of the Adam and Eve story of *Genesis*, that is, as a path leading back to the state of original unity with God.

30 In general, the Samaritan woman lives in the whole historical tradition of her people. For ex-

This causality of a new start in John 4 has obviously its counterpart in Kant's causality of freedom, but, in distinction, it is not reducible to an agent's (self's) "spontaneity" ("self-activity", B 446) without dependence on God.[31]

The causality of John 4, again, leads to a contradiction with respect to possible ways of God's presence, that is, between the immediate presence of God, independently of the contingencies of place and time, and the dependence of God's presence on some (contingent) place and time (see the fourth antinomy above).

Remark 2.1. *In terms of the system of mutually irreducible notions of possibility, imagination, and conception (as described by J.-Y. Béziau (2016)), we could interpret the somewhat surprising transition from the Samaritan woman's wish to get living water and eternal life to Jesus' request to her to call her husband, in the following way: we assume that her wish is* conceivable *for her (she understands its meaning), and also* imaginable *(for instance, in some Pentateuchal picture*[32]*), but there remains the question of* possibility *(aimed at by Jesus): she cannot receive eternal life while still being in causal dependence on the sensible temporal world (her past life). The solution is that she should make her wish independent of this temporal causal sequence by relating herself, in her self-knowledge (truth), to God. In application to Kant's antinomy: we could conceive (by categories) what it means to act out of freedom and imagine it in some intuitive form, but the possibility of such a causality opens up, for Kant, only in the intelligible moral world (otherwise,*

ample: "[...] our father Jacob, who gave us this cistern and drank from it himself with his children and his flocks" (John 4:12); and later: "Our ancestors worshiped on this mountain [Gerizim] [...]" (John 4:20). Cf. "The woman had found her security in the concept of *her* ancestors, but this is now transcended by the reference to '*the Father*' [...]. There is a progression from a very narrow and limited view of the exclusivity of one group to the liberating discovery that God is the Father of all [...]" (Botha (1991), p. 152). Sometimes, the determination by one's physical origin may be insurmountable: "Jesus himself testified that a prophet has no honor in his native place" (John 4:44).

31 For instance: "the idea of spontaneity, which could start to act from itself, without needing to be preceded by any other cause that in turn determines it to action according to the law of causal connection" (B 561); "a causality in our power of choice such that [...] it might [...] begin a series of occurrences *entirely from itself*" (B 563).
However, Kant points out: "The confirmation of the need of reason to appeal to a first beginning in the series of natural causes is clearly and visibly evident from the fact that (with the exception of the Epicurean school) all the philosophers of antiquity saw themselves as obliged to assume a *prime mover* for the explanation of motions in the world, i.e., a freely acting cause, which began this series of states first and of itself" (B 478).

32 E.g., Exodus 17:6: "Strike the rock, and the water will flow from it for the people to drink". Cf. Num 20:11.

the causality of freedom remains just ens rationis *and* ens imaginarium, *i.e., a concept and an intuition without an object, see B 347–349*).

(d) The fourth antinomy (dependence in existence). The solution confirms Thesis 4, according to which the necessary being, God, is immediately present (exists) – not in some space and time of a sensible world, but "in Spirit and truth": "the hour is coming, and is now here, when true worshipers will worship the Father in Spirit and truth" (John 4:23). In the sensible world, God is not immediately present, but probably only through a (long) intermediate sequence of the dependence on past events (coming, finally, to the beginnings of a religious tradition).

Here we observe the similarity with Kant's distinction between the "intelligible" necessity ("in Spirit and truth"), and "sensible" contingency. However, in John 4, not only is a necessarily existent being aimed at, but also its presence (in worship).[33] Thus the difference results between the presence of a necessary being in Spirit and truth, and a long "series of dependent existences" (B 587), eventually leading to some non-present (separate) necessary being.

We see that Spirit should be "something" that is independent of the contingent sensible world, although, at the same time, it should be somehow active in that world. It is also clear that the truth about the sensible world is not as such part of the sensible world, although it is related to that world. The question about *how* God can be immediately present to us in Spirit and truth remains open and is addressed in the final section of the dialogue.

2.3 The highest reality

We outline how the idea of God as the most real being (*ens realissimum*) (see (c) on page 62) can be traced back to the dialogue in John 4. In the Samaritan woman's utterance on what she knows about the worship of God in Spirit and truth, the idea of God as the source of the knowledge of "everything" (of truth) is clearly present:

> I know that the Messiah is coming, the one called the Anointed; when he comes, he will tell us everything. (John 4:25).

33 For Kant, in distinction, "the necessary being would have to be thought of as entirely outside the series of the world of sense (as an *ens extramundanum*), and merely intelligible; this is the only way of preventing it from being subjected to the law of the contingency and dependence of all appearances" (B 589).

The Messiah is "the one whom God has sent" and who "speaks the words of God" (John 3:34, cf. Dt 18:18,22). Thus the Messiah's knowledge of "everything" stems from God, who, in some way, possesses truth about "everything".[34] Since truth about everything is truth with respect to all possible predicates (according to the "principle of complete determination", see (c) on page 62), God obviously possesses the totality of all concepts (predicates). Here, a Kantian reasoning can be applied according to which God is completely determined by the idea of the "possession of all reality" (*Allbesitz der Realität*), as the entity having all positive properties (the negative ones being defined by means of the corresponding positive ones), and thus should be conceived as *ens realissimum* (B 604).

In distinction to Kant's view, in John 4 God is not merely a regulative ideal (or a postulated being), not even a being that will be present only in the future (as in the Samaritan woman's above-quoted statement, John 4:25), but is conceived as actually existing – present here and now, in the dialogue itself. This presence is indicated by the self-revealing words: "I am [*egō eimi*][35], the one who is speaking with you" (John 4:26).

3 Through dialogue to a metaphysical experience (a formalization)

Summarizing the preceding section, it can be said that in John 4 a metatheoretical conceptual structure is present that is analogous to Kant's system of transcendental ideas, one of the main differences being that these ideas should have actual denotation in religious experience instead of having a merely regulative or postulated role. Accordingly, the system of transcendental ideas should become a religiously based metaphysical system. This actual denotation of metaphysical

34 This "everything" is later reflected in the Samaritan woman's words: "Come see a man who told me everything I have done. Could he possibly be the Messiah?" (John 4:28, cf. John 4:39). The Samaritan woman's awareness that omniscience is grounded in God is confirmed by her wish to worship God after Jesus told her "everything she has done"; see the third antinomy above and John 4:18–20,29,39.

35 This is "an Old Testament self-designation of Yahweh" (New American Bible (2011), p. 1439, footnote ‡). "It [the affirmation of *egō eimi*] was used to manifest the living presence of God who makes himself accessible to his people. Jesus using this title manifests the visible presence of God to the Samaritan woman and eventually to the Samaritans" (Varghese (2009), p. 134).

concepts, as well as of concepts in general, should be grounded in their spiritual sense, as presented in John 4.[36]

In the following, we describe in a formal way how a religious reality of metaphysical concepts is achieved through the dialogical interaction of agents (Jesus, the Samaritan woman). To that end, we use a modification of justification logic (stemming from Gödel (1938); see also Artemov (2001)), where the concepts that were observed above as being transformed from metatheoretical to metaphysical ones will be interiorized into an object theory.

3.1 System QJDR

The vocabulary consists of individual constants $c, d, e, c_1, \ldots$ and individual variables $x, y, z, x_1, \ldots$; predicate letters P^i_j; $=$; $\neg, \rightarrow$; the quantifier symbol $\forall$; term operation symbols $+, \cdot, !$ and gen_x. Individual terms are individual variables and constants, and complex terms $(t+u)$, $(t \cdot u)$, $!t$, and $\mathsf{gen}_x(t)$, where t and u are individual terms.

Formulas are of the shape $Pt_1 \ldots t_n$, $t = u$, $\neg\phi$, $(\phi \rightarrow \psi)$, $\forall u\phi$, $t\colon \phi$, with the meaning 't has/gives evidence that ϕ', and $t\colon\colon \phi$ meaning 't wishes (requests) ϕ'. Symbols $\wedge$, $\vee$, $\leftrightarrow$ and $\exists$ are defined in the usual way.

We will also informally use accommodated English words to facilitate the understanding of translations. $\bot$ will stand for a contradiction, $\phi \wedge \neg\phi$.

The axiomatic system QJDR is designed on the basis of justification logic systems QLP by Fitting (2008) and FOLP by Artemov and Yavorskaya (2011), without factivity (an analogue of modal axiom **T**), with the addition of $=$ and wish operator $\colon\colon$, and with individual terms as evidence and wish terms.

The axioms are:

CPC	classical propositional tautologies
a	$\forall x\phi \rightarrow \phi(t/x)$, t is substitutable for x in ϕ
b	$\forall x(\phi \rightarrow \psi) \rightarrow (\phi \rightarrow \forall x\psi)$, $x \notin \mathsf{free}(\phi)$
Id	$x = x$
Rg	$x = y \rightarrow \exists z\, z\colon x = y \qquad \neg x = y \rightarrow \exists z\, z\colon \neg x = y$
Sub	$x = y \rightarrow (\phi(x) \rightarrow \phi(y))$, ϕ is atomic
JMon	$x\colon \phi \rightarrow (x+y)\colon \phi \qquad y\colon \phi \rightarrow (x+y)\colon \phi$

36 John 4 could be compared with John 3, where, in Jesus' dialogue with Nicodemus, we cannot see Nicodemus succeeding to access the spiritual reality (Nicodemus remained dependent on his naturalistic and historicist knowledge) (see Kovač (2015a)).

JK	$x: (\phi \to \psi) \to (y: \phi \to (x \cdot y): \psi)$
J4	$x: \phi \to\ !x: x: \phi$
J∀	$t: \phi \to \text{gen}_x(t): \forall x\phi,\ \ x \notin \text{free}(t)$
DMon, DK, D∀	like **JMon, JK, J∀**, respectively, with :: for :
DJK	$x:: (\phi \to \psi) \to (y: \phi \to (x \cdot y):: \psi)$
JDK	$x: (\phi \to \psi) \to (y:: \phi \to (x \cdot y):: \psi)$
DJ4	$x:: \phi \to\ !x: x:: \phi$

as well as the following special axioms:

SA1J	$InTruth(x) \to (x: \phi \to \phi)$
SA1D	$InTruth(x) \to \neg x:: \bot$
SA2	$InTruth(x) \to InTruth(!x)$
SA3	$InTruth(t) \to InTruth(\text{gen}_x(t))$, with $x \notin \text{free}(t)$
SA4	$(InTruth(x) \wedge InTruth(y)) \to InTruth(x \cdot y)$
SA5	$InTruth(x + y) \to (InTruth(x) \vee InTruth(y))$
SA6	*special axioms* including meaning postulates about non-logical symbols (to be introduced, sometimes implicitly, during the formalization of the dialogue of John 4).

Rules are modus ponens (**MP**), universal generalization (**UG**), and axiom justification (**AJ**): if $\vdash\ \phi$, then $\vdash\ c : \phi$, where ϕ is an axiom, and c a justification constant – according to some constant specification function **CS**, which assigns a justification constant to each axiom (in our translation of John 4, the assigned constants will be, informally, *j* and *s*).

In addition, some *facts* (possibly evidence and wishes) will appear as proof lines in the translation of the reasoning in the dialogue of John 4.

In some places with complex agents involving only one basic agent *t*, we will use an indexed expression $(t)^n$ for short.

Remark 3.1. *A unified definition of a model and a variable assignment can be proposed, with some basic features of the definition of a Mkrtychev model in* FOLP *(Fitting (2014); for propositional logic, see Mkrtychev (1997)), but extended with the identity relation and functions for wishes and requests, as well as including evidence and wish agents together with their complexes into the domain of a model (see Fitting's semantics of* QLP *for evidence in Fitting (2008), and the causal semantics for* QCGO *in Kovač (2015b)). Model,* $\mathfrak{M}$*, and variable assignment,* v*, make n-tuple* ⟨D*, I, v, Ev, Wish⟩*, where (a)* D* *is a set built on a basic set of individu-*

*als (*D*) and closed under evidence and wish operations,*[37] *(b)* I *is an interpretation function mapping individual constants to individual objects of* D*, and mapping relation symbols (predicate letters, term operation symbols) to relations on* D^**, (c)* v *is a variable assignment mapping each variable to a member of* D^**, so that now the denotation of a term,* $[\![t]\!]$ *(which is short for* $[\![t]\!]^{\mathfrak{M}}_{\mathrm{v}}$*), is* I(*c*) *or a complex object –* $\mathbf{!}[\![u]\!]$, $[\![u_1]\!]\mathbf{+}[\![u_2]\!]$, $[\![u_1]\!]\mathbf{\cdot}[\![u_2]\!]$, $\mathbf{gen}_x([\![u]\!])$ *– or* v(*x*)*, depending on whether t is a constant, a complex term, or a variable; finally, (d) evidence and wish functions* Ev *and* Wish *map each member of* D^* *to a subset of formulas, in analogy with the axioms:*

1. $\phi \in \mathrm{Ev}([\![t]\!])$ *or* $\phi \in \mathrm{Ev}([\![u]\!]) \Longrightarrow \phi \in \mathrm{Ev}([\![t]\!]\mathbf{+}[\![u]\!])$,
   $\phi \in \mathrm{Wish}([\![t]\!])$ *or* $\phi \in \mathrm{Ev}([\![u]\!]) \Longrightarrow \phi \in \mathrm{Wish}([\![t]\!]\mathbf{+}[\![u]\!])$,
2. $\phi \rightarrow \psi \in \mathrm{Ev}([\![t]\!])\ \&\ \phi \in \mathrm{Ev}([\![u]\!]) \Longrightarrow \psi \in \mathrm{Ev}([\![t]\!]\mathbf{\cdot}[\![u]\!])$,
   $\phi \rightarrow \psi \in \mathrm{Wish}([\![t]\!])\ \&\ \phi \in \mathrm{Wish}([\![u]\!]) \Longrightarrow \psi \in \mathrm{Wish}([\![t]\!]\mathbf{\cdot}[\![u]\!])$,
   $\phi \rightarrow \psi \in \mathrm{Ev}([\![t]\!])\ \&\ \phi \in \mathrm{Wish}([\![u]\!]) \Longrightarrow \psi \in \mathrm{Wish}([\![t]\!]\mathbf{\cdot}[\![u]\!])$,
   $\phi \rightarrow \psi \in \mathrm{Wish}([\![t]\!])\ \&\ \phi \in \mathrm{Ev}([\![u]\!]) \Longrightarrow \psi \in \mathrm{Wish}([\![t]\!]\mathbf{\cdot}[\![u]\!])$,
3. $\phi \in \mathrm{Ev}([\![t]\!]) \Longrightarrow t{:}\ \phi \in \mathrm{Ev}(\mathbf{!}[\![t]\!])$ *(positive introspection),*
   $\phi \in \mathrm{Wish}([\![t]\!]) \Longrightarrow t{::}\ \phi \in \mathrm{Ev}(\mathbf{!}[\![t]\!])$ *(positive introspection),*
4. $\phi \in \mathrm{Ev}([\![t]\!]) \Longrightarrow \forall x \phi \in \mathrm{Ev}(\mathbf{gen}_x([\![t]\!])),\ x \notin \mathsf{free}(t)$
   $\phi \in \mathrm{Wish}([\![t]\!]) \Longrightarrow \forall x \phi \in \mathrm{Wish}(\mathbf{gen}_x([\![t]\!])),\ x \notin \mathsf{free}(t)$.

The satisfaction of a formula is defined classically for atomic, compound and quantified formulas. For evidence and wish formulas, the conditions are as follows:

(*a*) $\mathfrak{M} \models_{\mathrm{v}} t{:}\ \phi \Leftrightarrow \phi \in \mathrm{Ev}([\![t]\!])$,
(*b*) $\mathfrak{M} \models_{\mathrm{v}} t{::}\ \phi \Leftrightarrow \phi \in \mathrm{Wish}([\![t]\!])$.

In principle, it should not be difficult to define the denotation of *InTruth* and of the non-logical relation symbols used below.

3.2 Translation of the dialogue

We will now translate some characteristic moments of the dialogue. Instead of formal descriptive symbols, we will mainly use abbreviations which by themselves indicate their meaning; *j* will denote Jesus, *s* the Samaritan woman, *w* physical (transient) water, and w' true (eternal) water. In the translation, we will focus on the interconnection of agents that is being established during the process of resolving contradictions and of increasing knowledge. Numbers in square brackets

37 In semantic metalanguage, the evidence and wish operations will be denoted by bolded evidence and wish operation symbols.

will indicate the reference of a reply or an opposition to a previous proposition in the dialogue.

Part 1

At the beginning of the conversation, Jesus actually asked for true water (w'), but this was not properly understood by the Samaritan woman. We will, first, show Jesus' request and beliefs (in a slightly simplified way) in the sense that the Samaritan woman thought they were meant, and thereafter we will express that they are being perceived so by the Samaritan woman:[38]

1	$j:: Gives(s, j, w)$	fact
2	$j: (j:: Gives(s, j, w) \rightarrow (\neg j:: Gives(s, j, w) \rightarrow \bot))$	AJ
3	$(j)^g: ((Jew(j) \wedge Samaritan(s)) \rightarrow \neg j:: Gives(s, j, w))$	SA6,AJ
4	$j: (Jew(j) \wedge Samaritan(s))$	fact
5	$((j)^g \cdot j): \neg j:: Gives(s, j, w)$	3, 4 JK, [1]
6	$((j \cdot j) \cdot ((j)^g \cdot j)):: \bot$	1, 2, 5 JK, JDK

As mentioned above, this is not quite what Jesus meant – it is how the Samaritan woman understood his words (the numerals in parentheses indicate the respective sentences of the formalization above):

7	$s: ((1) \wedge (2) \wedge (3) \wedge (4))$	fact
8	$(s)^h: (((1) \wedge (2) \wedge (3) \wedge (4)) \rightarrow (6))$	1-6 AJ
9	$((s)^h \cdot s): (6)$	7, 8 JK
10	$(s)^i: ((6) \rightarrow \neg InTruth(((j \cdot j) \cdot ((j)^g \cdot j)))$	SA1D, AJ
11	$((s)^i \cdot ((s)^h \cdot s)): \neg InTruth(((j \cdot j) \cdot ((j)^g \cdot j)))$	9, 10 JK
12	$((s)^j \cdot ((s)^i \cdot ((s)^h \cdot s))): \neg InTruth(j)$	9 SA4, AJ

We take that Jesus is aware of 12 (that the Samaritan woman, in her reasoning, does not believe him at the time). He therefore introduces another viewpoint by advancing his belief that what should happen is *Gives*(j, s, w'), referring to "true

38 For simplicity, the translation is in some places reductive on the ground of context. For example, 'to give a drink' is translated as 'to give water', and the Samaritan woman's question on how Jesus can ask her for a drink is translated simply as the assumption that Jesus is actually not asking her (because he should not ask her) to give him water. See a detailed interpretation, for example, in Botha (1991), pp. 115–122.

water". However, this is understood by the Samaritan woman again as $Gives(j, s, w)$, i.e., as referring to sensible water.

Let us now assume that $j: Gives(j, s, w')$ (1) as well as $s: \neg Gives(s, j, w)$ (2) hold. Since we take that j has evidence about all logical axioms, and since we thus get

$$j: (Gives(j, s, w') \rightarrow (Gives(j, s, w') \vee \neg Gives(s, j, w))),$$

so, starting from (1), $(j \cdot j): (Gives(j, s, w') \vee \neg Gives(s, j, w))$ follows. In an analogous way, starting from (2), we can derive $(s \cdot s): (Gives(j, s, w') \vee \neg Gives(s, j, w))$. From both sentences we obtain $((j \cdot j) + (s \cdot s)): (Gives(j, s, w') \vee \neg Gives(s, j, w))$ by **JMon**. Hence, we derive:

$$(j: Gives(j, s, w') \vee s: \neg Gives(s, j, w))$$
$$\rightarrow ((j \cdot j) + (s \cdot s)): (Gives(j, s, w') \vee \neg Gives(s, j, w)).$$

In addition, an application of the proof example by Artemov (2001), p. 10, to our context shows that evidence for $j: Gives(j, s, w') \vee s: Gives(j, s, w')$ includes the positive introspection of s and j, that is $(j: Gives(j, s, w') \vee s: Gives(j, s, w')) \rightarrow ((j \cdot !j) + (s \cdot !s)): (j: Gives(j, s, w') \vee s: Gives(j, s, w'))$.

Part 2

By a reasoning similar to that in Part 1, we obtain the following sequence: (1) $j:: Gives(j, s, w)$ (fact), (2) $(j)^k: (Gives(j, s, w) \rightarrow HasBucket(j))$ (**SA6, AJ**), (3) $((j)^k \cdot j):: HasBucket(j)$, (4) $j: \neg HasBucket(j)$ (fact), (5) $j: (HasBucket(j) \rightarrow (\neg HasBucket(j) \rightarrow \bot))$, (6) $((j \cdot ((j)^k \cdot j)) \cdot j):: \bot$. Like in Part 1, $((s)^{i'} \cdot ((s)^{h'} \cdot s)): \neg InTruth(((j \cdot ((j)^k \cdot j)) \cdot j))$ holds, and thus, for some complex built of s, an evidence for $\neg InTruth(j)$ again results (cf. **SA4**).

Therefore, Jesus introduces a more explicit distinction between w and w', which is noticed (although still not fully understood) by the Samaritan woman:

1	$j:: Gives(j, s, w')$	fact
2	$j: (Drinks(s, w') \rightarrow NeverThirsty(s))$, $j: (Neverthirsty(s) \rightarrow Drinks(s, w'))$	fact
3	$s:: Neverthirsty(s)$	fact
4	$(j \cdot s):: Drinks(s, w')$	2, 3 JDK
5	$s: (Drinks(s, w') \rightarrow Gives(j, s, w'))$	fact
6	$(s \cdot (j \cdot s)):: Gives(j, s, w')$	4, 5 JDK, [1]

However, the realizability of $(s\cdot(j\cdot s))::\ Gives(j, s, w')$, that is, of getting "a spring of water welling up to eternal life", depends on a further ascent, to be initiated by Jesus, in the Samaritan woman's knowledge.

Part 3

1	$s: j:: \exists x(Call(s,x) \wedge HusbandOf(x,s))$	fact
2	$s: \neg\exists x HusbandOf(x,s)$	fact, [1]
3	$j: HasDoneS(s)$	fact, "he told me everything I have done", John 4:39
4	$\forall y(HasDoneS(y) \rightarrow \neg\exists x HusbandOf(x,y))$	the meaning of *HasDoneS*, SA6
5	$(j)^{l}: (\neg\exists x HusbandOf(x,s) \wedge s: \neg\exists x HusbandOf(x,s))$	3, 4, assuming j: (2) as a fact, [2]
6	$s: j: HasDoneS(s)$	3 fact
7	$j: HasDoneS(s) \rightarrow Prophet(j)$	SA6
8	$s: (j: HasDoneS(s) \rightarrow Prophet(j))$	7 AJ
9	$(s\cdot s): Prophet(j)$	6, 8 JK
10	$\forall x(Prophet(x) \rightarrow InTruth(x))$	SA6
11	$(s\cdot s): (Prophet(j) \rightarrow InTruth(j))$	10 AJ*
12	$((s\cdot s)\cdot(s\cdot s)): InTruth(j)$	9, 11 JK, [12] of Part 1, cf. Part 2

(*In line 11, we apply **AJ** (for s) and $\forall$**a** to line 10, and hence, again by means of **AJ**, we get $s: (\forall x(Prophet(x) \rightarrow InTruth(x)) \rightarrow (Prophet(j) \rightarrow InTruth(j))$ as an intermediate step). Of course, the beliefs of lines 9 and 12 are true:

13	$Prophet(j)$	3, 7
14	$InTruth(j)$	10, 13

For 13, see Dt 18:18 and cf. with John 1.

Part 4

Thus, the Samaritan woman concludes that, in some place, God should be worshiped:

1	$(s)^l: (\exists x \mathit{Prophet}(x) \rightarrow \exists x \mathit{WorshipedIn}(god, x))$	SA6, AJ
2	$(s \cdot s): (\mathit{Prophet}(j) \rightarrow \exists x \mathit{Prophet}(x))$	∀a, tautology, AJ
3	$((s \cdot s) \cdot (s \cdot s)): \exists x \mathit{Prophet}(x)$	9 of Part 3 JK
4	$((s)^l \cdot ((s \cdot s) \cdot (s \cdot s))): \exists x \mathit{WorshipedIn}(god, x)$	1, 3 JK

However, there is controversy about the place of worship:

5	$\mathit{samaritans}: \mathit{WorshipedIn}(god, gerizim)$	fact, [4]
6	$\mathit{jews}: \mathit{WorshipedIn}(god, jerusalem)$	fact, [4]
7	$s: (\mathit{WorshipedIn}(god, gerizim) \rightarrow (\mathit{WorshipedIn}(god, jerusalem) \rightarrow \bot))$	fact
8	$((s \cdot \mathit{samaritans}) \cdot \mathit{jews}): \bot$	5–7 JK
9	$\neg \mathit{InTruth}(((s \cdot \mathit{samaritans}) \cdot \mathit{jews}))$	8 SA1J

According to the Samaritan woman it cannot be that both Samaritans and Jews are right regarding the place of worship. Thus Jesus points to worship in Spirit and truth as the solution to the controversy (see the next, final, part).

Final Part

1	$j: \forall x(\mathit{WorshipedIn}(god, x) \leftrightarrow (\mathit{InTruth}(x) \wedge \mathit{InSpirit}(x)))$	fact
2	$s: \exists x \mathit{TheAnointed}(x)$	fact ($\exists x$ referring also to the future), [1]
3	$s: \forall x(\mathit{TheAnointed}(x) \rightarrow (\mathit{InTruth}(x) \wedge \mathit{InSpirit}(x)))$	fact, [1] (cf.: x "will tell us everything", John 4:25)
4	$j: \forall x(\mathit{TheAnointed}(x) \rightarrow x = j)$	fact, [2]

We derive some consequences not explicitly stated in the Gospel text:

5	$(s)^n: (\forall x(\mathit{TheAnointed}(x) \rightarrow x = j) \rightarrow (\exists x \mathit{TheAnointed}(x) \rightarrow \mathit{TheAnointed}(j)))$	first-order logic, iterated JK
6	$((s)^n \cdot j): (\exists x \mathit{TheAnointed}(x) \rightarrow \mathit{TheAnointed}(j))$	4, 5 JK
7	$(((s)^n \cdot j) \cdot s): \mathit{TheAnointed}(j)$	2, 6 JK
8	$((s \cdot s) \cdot (((s)^n \cdot j) \cdot s)): (\mathit{InTruth}(j) \wedge \mathit{InSpirit}(j))$	3, 7 JK*
9	$((j)^m \cdot ((s \cdot s) \cdot (((s)^n \cdot j) \cdot s))): \mathit{WorshipedIn}(god, j)$	1, 8 JK*, [1]

(*See analogous note * for line 11 of Part 3.) Line 9 includes some propositional logic as evident to j. Of course, according to **J4** the following is derivable: $!(((s)^n \cdot j) \cdot s): (((s)^n \cdot j) \cdot s): \mathit{TheAnointed}(j)$ as well as $!((j)^m \cdot ((s \cdot s) \cdot (((s)^n \cdot j) \cdot s)): ((j)^m \cdot$

$((s \cdot s) \cdot (((s)^n \cdot j) \cdot s))$: *WorshipedIn(god, j)*. In addition, everything that was in the proof until line 9 stated for *s* holds for *j*, too, from where and from line 14 of Part 3, it follows (by **SA1J**):

10	$\forall x(TheAnointed(x) \to x = j)$	see 4
11	$InSpirit(j)$	see 8
12	$WorshipedIn(god, j)$	see 9

Thus, non-formally expressed, the result is that Jesus (*j*) is a true, spiritual self ("I am") in which God should be worshiped.

As the Samaritan woman returned to her town, she told the people about what had happened; thus she may have reasoned in more general terms. For example, her reasoning from Part 2 might also have been as follows: from her generalized wish

$$s :: (HumanInTown(x) \to Neverthirsty(x)),$$

and from j: $(Neverthirsty(x) \to Drinks(x, w'))$ (cf. Part 2, line 2) it follows (with a bit of *s*'s evidence of propositional logic) that

$$((s \cdot s) \cdot j) :: (HumanInTown(x) \to Drinks(x, w')).$$

Obviously, from

$$s : (Drinks(x, w') \to Gives(j, x, w'))$$

we can derive (again, on the ground of *s*'s evidence of propositional reasoning)

$$((s \cdot ((s \cdot s) \cdot j)) \cdot s) :: (HumanInTown(x) \to Gives(j, x, w')).$$

Now, according to **D**$\forall$, we conclude:

$$\mathbf{gen}_x(((s \cdot ((s \cdot s) \cdot j)) \cdot s)) :: \forall x(HumanInTown(x) \to Gives(j, x, w')).$$

In distinction to John 3 (a conversation with Nicodemus), we see how through the complexity of the agents' interconnections and during the progress of resolving contradictions and of gaining knowledge, the agents become "reborn" in the "water" of the dialogue and "in Spirit".

At the same time, we see how metaphysical concepts, especially the concepts of self, world and God, if approached from a religious viewpoint, could obtain a sort of objective reality beyond the realm of sensible objects.

Acknowledgement

I am grateful to Mirosław Szatkowski for the invitation to the conference *God, Time, Infinity*, Warsaw, September 22–24, 2015, where the first version of the paper was presented. This work has been supported by the Croatian Science Foundation under project IP-2014-09-9378.

Bibliography

Achourioti, T. and van Lambalgen, M. (2011), "A Formalization of Kant's Transcendental Logic", in *Review of Symbolic Logic*: 4, 254–289.

Artemov, S. (2001), "Explicit Provability and Constructive Semantics", in *Bulletin of Symbolic Logic*: 7, 1–36.

Artemov, S. and Yavorskaya (Sidon), T. (2011), "First-order Logic of Proofs", in *CUNY Ph.D. Program in Computer Science*, TR-2011005.

Béziau, J.-Y. (2016), "Possibility, Imagination and Conception", in *Principios*: 23, 59–95.

Botha, J. E. (1991), *Jesus and the Samaritan Woman: A Speech-Act Reading of John 4:1-42*, Leiden: E.J.Brill.

Fitting, M. (2008), "A Quantified Logic of Evidence", in *Annals of Pure and Applied Logic*: 152, 67–83.

Fitting, M. (2014), "Possible World Semantics for First-Order LP", in *Annals of Pure and Applied Logic*: 165, 225–240.

Gödel, K. (1947/1964), "Cantor's Continuum Problem", in *Kurt Gödel, Collected Works*, Vol. 2, edited by S. Feferman *et al.*, Oxford: Oxford University Press, 1990, 254–270.

Gödel, K. (1938), "Vortrag bei Zilsel", in *Kurt Gödel, Collected Works*, Vol. 3, edited by S. Feferman *et al.*, Oxford: Oxford University Press, 1995, 86–113.

Jojko, B. (2012), *Worshiping the Father in Spirit and Truth*, Rome: Gregorian and Biblical Press.

Kant, I. (1910–), *Kant's Gesammelte Schriften*, edited by the Königlich Preussische Akademie der Wissenschaften, Berlin: Reimer; and Berlin and Leipzig: de Gruyter. Also *Kants Werke*, I–IX, Berlin: de Gruyter, 1968.

Kant, I. (1781/1787) *Kritik der reinen Vernunft*, in Kant (1910–), Vol. III–IV, 1st ed. 1781 (A), and 2nd ed. 1787 (B).

Kant, I. (1783) *Prolegomena zu einer jeden künftigen Metaphysik*, in Kant (1910–), Vol. IV.

Kant, I. (1998) *Critique of Pure Reason*, translated and edited by P. Guyer and A. W. Wood, Cambridge [UK]: Cambridge University Press.

Kant, I. (1997) *Prolegomena to Any Future Metaphysics*, translated by P. Carus, Chicago, La Salle: Open Court.

Kovač, S. (2011), "Logika i vjera", in *Suvremena znanost i vjera : Contemporary Science and Faith*, edited by Z. Primorac, Mostar, Ljubljana, etc.: Fakultet prirodoslovno-matematičkih i odgojnih znanosti Sveučilišta u Mostaru, 69–84.

Kovač, S. (2015a) "Logic and Truth in Religious Belief", in *God, Truth, and Other Enigmas*, edited by M. Szatkowski, Berlin, etc.: de Gruyter, 119–132.

Kovač, S. (2015b) "Causal Interpretation of Gödel's Ontological Proof", in *Gödel's Ontological Argument: History, Modifications, and Controversies*, edited by K. Świętorzecka, Warsaw: Semper, 163–201.

Mkrtychev, A. (1997), "Models for the Logic of Proofs", in *Logical Foundations of Computer Science*, edited by S. Adlan and A. Nerode, Berlin, Heidelberg, etc.: Springer, 266–275.

Nestle-Aland Novum Testamentum Graece (2012), Stuttgart: Deutsche Bibelgesellschaft.

New American Bible, Revised Edition (2011), Charlotte, NC: Saint Benedict Press. Also: *New American Bible*, 2002, http://www.vatican.va/archive/ ENG0839/_INDEX.HTM

Novakovic, L. (2013), "Jews and Samaritans", in *The World of the New Testament*, edited by J. B. Green and L. M. McDonald, Grand Rapids: Baker Academic, 207–216.

Tiles, M. (2004), "Kant's Logic", in *Handbook of the History of Logic*, Vol. 3, edited by D. Gabbay and J. Woods, Amsterdam etc.: Elsevier, 85–130.

Varghese, J. (2009), *The Imagery of Love in the Gospel of John*, Rome: Gregorian and Biblical Press.

Zbigniew Król

Basic Intuitions Concerning the Concept of Infinity in Mathematics from a Historical and Theological Point of View

Thus the conquest of actual infinity may be considered an expansion of our scientific horizon no less revolutionary than the Copernican system or than the theory of relativity, or even of quantum and nuclear physics.
Fraenkel and Levy (1976), p. 40.

1 Introductory remarks

The history of human understanding of itself and reality consists in many anonymous, however, fundamental discoveries. There had to have been someone who noticed that there were such "things" as time, space or infinity. It is impossible to experience them through the senses – sight, hearing, smell or touch. These remarkable insights are works of unknown geniuses who still guide and determine our everyday, as well as scientific, life.

The concept and problem of infinity is still far from having been fully analyzed. However, we know much more about infinity now. The intuition of infinity consists of many factors which determine the possible directions of further analysis. We have the intuitions of infinitely large and small quantities,[1] actual and potential infinity, intuitions of infinite series and infinitely divisible quantities, as well as intuitions of many actually infinite objects: infinite straight line, space, surface, the set of natural numbers etc. We can imagine infinity in space and time, e.g. the idea of eternity, or in perfection. Such ideas, in a natural way, are directed toward the concept of God, who is often considered as the source and cause of every kind of "human" infinity. One can say much more: the analyses of the consistency of the concept of God, who is considered as omnipotent, infinitely powerful, free, eternal etc., are the main sources for the development of the concept of infinity, not only from the historical point of view.

1 Cf. for instance non-standard analysis; see Robinson (1966).

https://doi.org/9783110594164-006

2 The development of the concept of infinity

The ways of understanding the concept of infinity have been changing through the ages. The concept was, at first, interpreted with the help of the concept of a limit (*peras*) and it was defined as *apeiron*, i.e. non-limited, unlimited, undetermined etc. Something was infinite when it lacked an ascertainable limit. Thus, infinite series are infinite because there is no limit in increasing them infinitely. It seems that the first kind of analyzed infinity was potential infinity. In the same way, lines are infinite because one can divide them without any limits, i.e. any part of the divided continuous quantity can be divided later into smaller and smaller parts, or one can produce them without any limits.

This type of infinity belongs to the type of undefined, undetermined quantities. Therefore, Plato, in his unwritten doctrine, *Philebus* or *Parmenides*, considered that everything is composed of two factors: the two Highest Principles, i.e. the One and the (Undetermined), the Dyad of that which is Great and Small. The One imposes some limits on the Dyad. The highest principles are also present in ancient mathematics. For instance, every mathematical entity in Euclid's *Elements* is composed of some limiting factors and of something undetermined; see, the lines or figures in *Elements*. The lines without endpoints were called "infinite". However, the lines or figures are finite but of undefined size. They play the role of geometrical variables; see also the opinion of Euler in his (Euler, 1797).[2]

As I argue in my book, *Platonism and the Development of Mathematics. Infinity and Geometry* (Król (2015)) actually infinite objects were actually known in Antiquity. However, they were excluded from mathematics and geometry. For instance, there is no infinite line, surface or space in Euclid's *Elements* or by Archimedes. The problem of operating with the wholes (concepts) of actually infinite scopes is the main reason why the ancient authors did not consider the problem of a possible change of the basic line in *Book X* of *Elements*.

The only two exceptions to ancient finitism were:

1. Proclus' proof that to a given line there is only one (infinite) parallel line different from the given line;[3]

---

2 Cf. also Euler (1990).

3 Only this first example demonstrates the unique antique use of actual infinity in a mathematical proof. Proclus, explaining theorem I.30, i.e., that straight lines parallel to the same straight line are also parallel to one another, says (contrary to Euclid): "For we must think of parallel lines as produced indefinitely, and ***AH*** when produced coincides with ***HB***; it is therefore the same as it, and not another line. Therefore all the parts of a parallel line are themselves parallel to the straight line to which it is parallel, both to the whole of it and to its parts. Thus ***AH*** is parallel to

2. an attitude taken by Apollonius in his *Conics*.[4] (This last case is not as evident as the first one.)

It is possible to indicate basic historical changes in the understanding of the concept of infinity. At the very beginning, infinity is considered as potential infinity and it is analyzed with the use of the concept of "unlimitedness". Some other concepts or intuitive convictions are also essential. For instance, that the whole is greater than a part of it or that the sum of infinitely many quantities which are "greater than nothing" is infinitely great. There are also some beliefs which are non-verbal, i.e., that there is only one infinity in every kind.

3 God's point of view and the emergence of actual infinity in mathematics

The so-called, *God's point of view* was very important in the process of the emergence of actual infinity in mathematics. Philosophers started their analysis from *God's point of view*, i.e. they considered logical situations in which the assumptions concerned the omnipotent, eternal and all-knowing subject of cognition. For a limited human being, it was impossible to actually divide a continuum to infinity during a finite lifetime etc. However, it was possible for him to analyze the

KD and ***HB*** to ***CK***; for when produced indefinitely they remain nonsecant." (Cf. Proclus (1987), p. 294). He also proves the uniqueness of the parallel line through the given point and parallel to the given line: "From the same point two perpendiculars cannot be drawn to the same straight line, nor through the same point can two parallels be drawn to the same straight lines." (Proclus (1987), p. 296). However, Proclus' statement is in contradiction with Euclid's explicit proofs from *Elements*. Proclus mixes the meanings of the actually infinite line and the undefined or undetermined line: there is really only one parallel and undefined line; i.e., a line without limited points, and many finite parallel lines. He is also not firm in the above statement, and contradicts himself in many places; cf. p. 303 ("For a parallelogram is formed by the equal and parallel lines") or p. 312.

4 In contradiction to Euclid, Archimedes and others who, in their definitions and propositions, use only determined finite cones, Apollonius defines an indefinite conic line and conic surface, and discerns these objects from every given cone: "If a straight line indefinite in length, and passing through a fixed point, be made to move round the circumference of a circle which is not in the same plane with the point, so as to pass successively through every point of that circumference, the moving straight line will trace out the surface of a double cone, or two similar cones lying in opposite directions and meeting in the fixed point, which is the apex of each cone." (Heath (1896), p. 1). Based on Apollonius' *Conics*, it is necessary to differentiate between the actually infinite line or surface and undefined in length, i.e. an unlimited line and surface respectively.

logical consequences of such premises as if a man could exist like God. This last point is exemplified by the medieval discussions concerning the concept of God's omnipotence or some properties of the continuum. Without the consideration of God's properties and divine possibilities or without the application of God's cognitive perspective, the use of actually infinite objects in mathematics would not have been possible.

One of the most well-known and broadly discussed argument in the Middle Ages was Henry of Harclay's,[5] who lived a little later than Nicole Oresme. The case is reported by William of Alnwick in his *Determinationes*, next quoted by Adam Wodeham in *Tractatus de indivisibilibus*:[6]

> God actually sees or knows the first beginning point of a line, and any other point which it is possible to pick out in the same line. Therefore, either [i] God sees that, in between this beginning point of the line and any other point in the same line, a line can intervene, or [ii] not. If not [i.e., (ii)], then he sees point immediate to point, which is what we propose. If so [i.e., (i)], then, since it is possible to assign points in the intermediate line, those points will not be seen by God, which is false. This consequence is clear, for according to what we have posited a line falls between the first point and any other point (of the same line) seen by God, and consequently, there is some midpoint between this point and any other point seen by God. Therefore this midpoint is not seen by God.

Three main groups of theories concerning a continuum were formulated in the 14th century as a consequence of such discussions: 1. there are no points at all being the parts of a continuum, 2. a continuum is composed of infinitely many parts, each of which is also a continuum; an example are the views of Gregory of Rimini.[7] A continuum is composed of infinitely many indivisible parts, for example, points. More information about this topic and about the emergence of actual infinity in mathematics can be found in my book.

The works of Nicole Oresme (c. 1320–1387) are the most important from the *mathematical* point of view in this process.[8] We find in his works a very mature and conscious use of the concept of actual infinity, both in geometry and in arithmetic. Nicole Oresme applies the infinite concepts in these two domains. The first use is exemplified by the summation of some infinite series. The second attitude can be found in his theorems concerning the commensurability and incommensurability of circular motions. However, the geometrical way of thinking is also present in arithmetic because he considers arithmetical theorems mainly from the geomet-

---

5 The reader can find more examples in Grant (1982).

6 See, Cross (1998), p. 89.

7 Analysis and literature on the subject, see, Cross (1998), p. 3.

8 In this point, I adopt the results of my considerations from Król (2015), Chapter 13.

rical point of view. This attitude is exceptional because, at almost the same time, purely logical and arithmetical reasoning are present, for instance, in the works of the Oxford's *calculatores*. In Oresme's works, one can find for the first time the prototypes of infinite lines, surfaces and also infinite, three-dimensional space; see, Król (2015), Chapter 13.

The word "summation" of some infinite series of numbers can be misleading because, in reality, Oresme grounds the summation on a *division* of the given *finite* magnitudes into an *actually* infinite number of proportional parts. Oresme's main mathematical innovation is the adoption of God's point of view: he performs the aforementioned divisions and proofs, assuming that such a division is made *actually* in infinity. This small step is absolutely revolutionary from the mathematical point of view.

The last six questions of his work *Tractatus de configurationibus qualitatum et motuum*, III.viii–III.xiii, contain constructions and proofs with the use of actually infinite geometrical objects and infinite series. Oresme applies the actually infinite series only for geometrical problems and they correspond to some easily imaginable geometrical objects. For instance, in question III.viii, he considers the following problem: "A finite surface can be made as long as we wish, or as high, by varying the extension without increasing the size."[9] In order to demonstrate how it is possible also in the case of a surface, he considers two one-square foot identical squares. He divides both of them into *actually* infinitely many proportional parts (sub-surfaces have areas equal to $1/2, 1/4, 1/8, \dots$, which are designated by Oresme as ***E***, ***F***, ***G***, etc.), the sum of which is – in an easily visible way – equal to the one foot square in both cases. Then, he uses the actually infinite number of these parts of the second square and stands them on top of the proportional parts of the first square: first he places $1/2$ of the second square on top of the first one, then, $1/4$ of the remaining half of the second square he places on top of the previous $1/2$, etc., and all the parts have one collinear side which extends to infinity. He also demonstrates that the area of such a "stepped" actually to the infinity figure which actually contains infinitely many parts of determined, finite areas, is also finite, i.e. it is equal to a two square foot square. From the modern point of view, Oresme geometrically sums up the actually infinite series of the form: $1 + 1/2 + 1/4 + \dots + 1/(2n) + \dots = 4(1/2) = 2$. He comments:

> Then upon this whole let the second part, namely ***F***, be placed, and again upon the whole let the third part, namely ***G***, be placed, and so on for the others to infinity. When this has

9 Cf. Clagett (1968) , p. 413.

> been done ("Quo facto"), let the base line ***AB*** be imagined as being divided into parts continually proportional according to the ratio of 2 to 1.... (Clagett (1968) p. 415)

This very simple step is revolutionary, "when this has been done", enables the strict analysis of *actually* infinite geometrical objects. The summations, in every case (cf. the next questions of Part III De *configurationibus...*) are based on some divisions into actually infinitely many proportional parts of some *finite* magnitudes.

As I demonstrate in my book, the discovery and use of actually infinite objects in mathematics is a very complicated historical process which reaches its *akmé* in the works of Nicole Oresme, Cavalieri, Torricelli, Newton and Euler. However, Georg Cantor said much more about the nature of infinite collections than anybody before him.

Infinite objects in mathematics, especially in geometry, were in common use in his times as a result of the application of God's point of view. However, infinity, even in Cantor's times, was considered as God's own essential property and as a quality which can help to discern created things from eternal God. Actual infinity has been reserved for God itself in Christian theology since the Middle Ages and even since ancient times.[10]

It is a well-known fact that Cantor believed that the theory of transfinite numbers was revealed to him by God. Cantor was aware of the philosophical implications of this theory. He speaks about it in the Introduction to *Grundlagen einer allgemeinen Mannigfaltigkeitslehre*. Opposition to *Cantor's Paradise* among mathematicians (Kronecker, Brouwer, Poincaré) interplays with the voices of opponents arguing from the theological point of view (Neo-Thomists, cardinal Johannes Franzelin).[11] The presence of actually infinite collections and other infinite objects can be reconciled with traditional Christian theology only if the objects of mathematics are some fictions or are "imaginary" in the sense of old medieval theories.

In the same way and because of theological obstacles, Descartes resigned from the concept of actual infinity in mathematics; cf. Król (2015).

Descartes maintains the ancient difference between "infinite" and "indeterminate" things. "Infinite", in all aspects, is only God because He is "perfect", i.e. there is no possibility to be completed (in an aspect) for Him. The other things can be completed and, therefore, they are "indeterminate" and not perfect. Descartes

10 Cf. Davenport (1997), pp. 263–295.

11 The reader can find more information together with the relevant sources in Dauben (1977, 1979, 2004).

explains the difference between the terms *infinitus* and *indeterminatus* in his *Principles of Philosophy, Chapters XXVI*. However, he adopts God's point of view in many places and considers actually infinite objects as well as collections of points; cf. Król (2015), Chapter 14.2. It means that his methods are Platonic.

4 Infinity in modern set theory

From this perspective, it is interesting to analyze what *God's point of view* in modern set theory relies on, i.e., what are Platonic methods in reference to the concept of infinity. In the description of the methods, I omit a rather well-known point concerning the non-predicativity ("circularity") of the axiom of infinity.

In modern mathematics, the problem of infinity in set theory seems to be one of the most important. There are two basic strategies of the construction of an infinite set: upward and downward. The first is usually applied in a universe of well-founded sets and creates an infinite "tower" of sets under the roof of an infinite limit-set. The downward strategy enables the creation of non-well-founded infinite sets, e.g., a Mirimanoff-like infinite series of nested sets located inside a "bottomless pit".

According to the first strategy, which is usually formalized within the frames of **ZF(C)** or **NBG** etc., the axiom of infinity can have many formulations. However, the basic ones are as follows:

A.1. $\exists z[(\exists x \in z).\forall y.\neg(y \in x) \wedge (\forall x \in z)(\exists y \in z).x \in y]$;

A.2. $\exists z[(0 \in z) \wedge \forall x.(x \in z \rightarrow \{x\} \in z)]$;[12]

A.3. $\exists z[(0 \in z) \wedge \forall x.(x \in z) \rightarrow (x \cup \{x\}) \in z]$;[13]

A.4. $\exists z[(0 \in z) \wedge (\forall xy.(x \in z \wedge y \in z \rightarrow x \cup \{y\} \in z)]$;

A.5. $\exists z \exists x.(x \subset z \wedge x \neq z \wedge x \div z)$; (there is a reflexive set, "$x \div z$" – "there is a one-to-one correspondence between sets x and z";[14]

A.6. There is an infinite set (in Bernays' sense);[15]

A.7. $\exists z \neq 0[\forall x.(x \in z \rightarrow \exists y.y \in z \wedge x \subset y \wedge x \neq y)$.[16]

12 Cf. Zermelo (1908), pp. 261–281.

13 Cf. Neumann (1923), pp. 199–208.

14 This is, obviously, Dedekind's definition.

15 Cf. Bernays (1958), p. 150. Bernays uses the formal statement which asserts that the class of natural numbers is a set, i.e., "is representable" in his wording. From this and Definition 1 follows the existence of an infinite set.

16 Cf. Bernays (1958), p. 150. I have slightly modified the original version of the axiom which was originally introduced by J. von Neumann and K. Gödel.

On the other hand, one can develop many concepts, such as the concepts of equinumerosity, ordinal number, transfinite induction, natural number, numeral or finite set, all of these without any concept of an infinite set.[17] Therefore, the next possibility to grasp the concept of infinity in set theory is to define an infinite set as a set which is not finite.

Def.1. $\mathrm{Inf}(z) \equiv \neg\mathrm{Fin}(z)$

In the system of Bernays, the existence of an infinite set follows from axiom **A.6**.

It appears that the Zermelo's axiom **A.2** follows from **A.7**. In fact, these axioms are equivalent; cf. Bernays (1958), pp. 148–150. Moreover, all the above axioms are equivalent to one another [18] on the basis of some other axioms of **ZF**, i.e. the axioms of Pairing, Union, Power-Set, the axiom schema of Subsets and Replacement but without the axioms of *Foundation and Choice*.[19] We also know that the axioms of infinity are independent from the other axioms of set theory **ZF(C)**.

The above axioms and definitions of an infinite set are non-predicative, i.e., they are circular.[20] From the technical point of view, it means that we cannot construct an actually infinite set from the previously given finite sets or prove its existence from other axioms. One can simply postulate and accept the existence of an infinite set.

Obviously, many other versions of the axiom of infinity are possible. The idea in constructing them is to postulate the existence of a set which, together with all of its elements, also contains a next "bigger" or "separate" (different) element. The other axioms can produce a *potentially* infinite number of different elements. However, they cannot secure the existence of a set which contains all elements which are constructed with the use of a specific method. For instance, one can build a potentially infinite series of different sets $x, P(x), P(P(x)), P(P(P(x))), \ldots,$ with the use of the Power set axiom, but the existence of a set containing all of them does not follow from the other axioms of set theory. Therefore, we need an axiom of the form:

A.8. $\exists z \exists x \neq 0[(x \in z) \wedge \forall x(x \in z \rightarrow P(x) \in z)]$.[21]

17 Cf. Bernays (1958).

18 **A.1** is equivalent to **A.2** in **ZF** because the only atom in **ZF** is the empty set 0. A quite different situation is in **ZFA** or in **ZFC**$^-$ + **AFA**.

19 Cf. Fraenkel, Bar-Hillel and Levy (1973), pp. 45–47.

20 For more details, see, Król (2006).

21 Obviously, when there is already another axiom of infinity, it is easy to prove **A.8**.

One can use any method which can create a potentially infinite series of different sets. It is also possible to postulate simply that there is a potentially infinite number of such different sets. One can, or not, use the empty set or other atoms (in **ZFA**, i.e., **ZF** with atoms). However, not in every case can such an axiom be used for the construction of the set of natural numbers; see also below. Some examples of such axioms are given below:

A.9. $\exists z\,[(0 \in z) \wedge (\forall x \in z\, \exists y \in z(y \neq 0 \wedge (x \cap y) = 0 \wedge (x \cup y) \in z)]$;
A.10. $\exists z \exists x[(x \in z \wedge x \neq 0) \wedge (\forall c \in z\, \exists y \in z\,(y \neq 0 \wedge (c \cap y) = 0 \wedge (c \cup y) \in z)]$;
A.11. $\exists z \exists x[(0 \in z \wedge x \in z \wedge x \neq 0) \wedge (\forall c \in z\, \exists y \in z\,(y \neq 0 \wedge c \neq y \wedge (c \cup y) \in z)]$;
A.12. $\exists z \neq 0[\forall x \in z\, \exists y \in z\,(y \neq 0 \wedge (x \cap y) = 0 \wedge (x \cup y) \in z)]$.

The most simple axiom of infinity in this group is **A.12** and it enables one to use not only the empty set $x = 0$.

The following questions arise: is it true that every infinite set is well-founded or isomorphic to an ordinal number? Are all well-founded axioms of infinity equivalent? What are the logical relations between the concepts of actual infinity, equinumerosity and well-foundedness? Is there only one Cantor's Paradise possible? Does every infinite set belong to Cantor's Paradise?

Some of the above axioms assert the existence of an infinite set only if the universe of sets is well-founded. For instance, if the set "$z \in z$" is not excluded then, for example in **A.1** or **A.12**, set z can be finite for $z = \{x, z\}$. Nevertheless, the usual axioms of infinity in **ZFC**, such as **A.2–A.3** work well also in non-well-founded set theories; cf. for instance the (most) systems of non-well-founded set theories described in the book of P. Aczel, for instance in **ZFC**$^{-}$**+AFA.**[22] From the above facts one can infer that not every axiom **A.1–A.12** is equivalent to the other axioms in *every* formal environment.

Are there infinite sets which are non-well-founded? We are now going to consider some examples of infinite sets according to the above-mentioned second strategy of the construction of infinite sets. We can present some examples of such axioms. The first axiom concerns the case when the set "$x \in x$" as an element of z is not excluded:

A.13. $\exists z \exists x(x \in x \wedge x \in z) \wedge [\forall x \in z\, \exists y \in z\,(y \neq 0 \wedge (x \cap y) = 0 \wedge (x \cup y) \in z)]$.

(From **A.13** follows that $x \neq z$, $y \neq x$, $y \neq z$.) Obviously, there are other possible variants of this axiom depending on if the null set belongs to z or x.

22 Cf. Aczel (1988).

When "$x \in x$" is excluded from z, we have, for instance (a "Mirimanoff set"):

A.14. $\exists z \neq 0 \forall x.\{[x \in z \rightarrow \neg(x \in x)] \wedge \exists x \forall y\ (x \in y \wedge y \in z \rightarrow x \in z) \wedge \forall x \in z\, \exists y\ (y \in x \wedge y \neq 0)\}$.

A.14.1. $\exists z \neq 0\ \{\forall x, m \in z.\neg(x \in x) \wedge (x \in m \wedge m \in z \rightarrow x \in z) \wedge [\forall x \in z\, \exists y\ (y \in x \wedge y \neq 0)]\}$.[23]

The last two sets have no $\in$-minimal elements and all the finite, as well as infinite, sequences of the form "$x \in m \in u \ldots \in x$" are excluded.

In these last cases, one can obtain a "two-sided" infinite set, i.e. a set which is infinite "up and down":

A.15. $\exists z \neq 0\, \forall x \in z.\neg(x \in x) \wedge \exists y \in z\ (x \in y \wedge y \neq z) \wedge \forall x \in z\, \exists y \in z.$ $(y \in x \wedge y \neq 0)$.

The last formula indicates other possible axioms in the case of well-founded sets:

A.16. $\exists z \neq 0\, \forall x \in z\, \exists y \in z\ (x \in y \wedge y \in z)$;
A.17. $\exists z.0 \in z \wedge \forall x \in z\, \exists y \in z\ (x \in y \wedge y \in z)$.

The next group of non-well-founded axioms of infinity contains axioms which provides a set with an infinite **circular** sequence of the forms "$z \in m \in u \in n \ldots \in z$" or "$z \in m \in u \in n \ldots \in u \in z$", etc. However, such sets are not useful for a definition of natural numbers.

Such axioms as **A.16** and **A.17** are not necessarily transitive. On the other hand, **A.14** (or **A14.1**) are transitive. However, they can contain many finite, as well as infinite, branches. Therefore, it is difficult to control their cardinality and to construct from them a set of natural numbers. One can easily reformulate such axioms and impose the condition of transitivity or a condition eliminating undesired branches; cf. below.

The next problem is how one can enable the construction of ordinal numbers and **N** in a non-well-founded, as well as well-founded, case. At first, it is necessary to discern a unique specific element in such a set. We usually postulate that the null-set 0 (or an atom; cf. **A.1**) is an element of an infinite set. The presence of such a discerned element makes it possible to find a minimal set, i.e. a set which is contained in every infinite set.[24] For example, using **A.2** or **A.3**, one cannot be sure if there are no other elements a, like 0, from which it is possible to start one more

23 Obviously, the last two axioms are equivalent.
24 One can prove, assuming the axiom schema of replacement, that the union of every family of *sets* is also a set.

and different infinite sequence of sets, e.g. $a, \{a\}, \{\{a\}\}, \ldots$. The construction of a minimal set excludes such an unwanted series.

Let us consider, for instance, how one can construct a set of natural numbers with the use of **A.14.1**. First, it is necessary to eliminate other branches from the set z, for instance:

TR. $\forall x \in z \forall y \in z. x \subset y \vee y \subset x$; $(\forall x \in z \forall y \in z. x \in y \vee y \in x)$.

TR excludes other "branches" in the set z from **A.14.1** and we obtain a singular infinite "thread" of the subsets of z. We also need a condition which asserts that between any two elements x, y of z, there is only a well-founded sequence of sets $m, n, k, \ldots$, such that $x \in m \in n \in k \in \ldots \in y$. For example:

Suc. $\exists w \in z. \neg(\exists c \in z. w \in c) \wedge \forall x \in z \exists s \in z. (x \in s \wedge \neg \exists m \in z. x \in m \wedge m \in s)$.

It is easy to demonstrate that such w is only one in z. Let $w = \text{Suc}(z)$ and $\text{Suc}(x \in w) = s \in z$ that $\neg \exists m \in z. x \in m \wedge m \in s$:

$\mathbf{Suc}(x)_z = s$ iff $s \in z \wedge x \in z \wedge \forall m \in z. \neg(x \in m \wedge m \in s)$.

Our axiom of infinity is: **A.14.1** $\wedge$ **Tr** $\wedge$ **Suc** (with obvious bracketing). Next, it is easy to define: $0 = z$, $S(w) = 1 = S(\text{Suc}(z))$, $n + 1 = S(S(n)) \ldots$, etc.

One more possibility is to define an initial segment of x in z from **A.14.1** $\wedge$ **Tr** $\wedge$ **Suc:**

$\mathbf{Ext}(x)_z = \{s \in z : x \in s\}$.

For every x in z, $\mathbf{Ext}(x)_z$ is well-founded and $\mathbf{Ext}(x)_z = \mathbf{Ext}(y)_z$ iff $x = y$. Therefore, there is a sequence of ordinal numbers in Cantor's Paradise which is isomorphic to the sequence of the well-founded sets $\mathbf{Ext}(x)_z$, $x \in z$, ordered by the usual relation $<$. One can also use these last sets and define elements of **N**. There is also the set **N** of all initial segments of x in z.

Thus, there are at least two possibilities to define natural numbers with the use of the non-well-founded axiom of infinity **A.14.1** $\wedge$ **Tr** $\wedge$ **Suc**, the axiom of separation and replacement:

1. it is possible to construct **N** from z; e.g., from $Z = \{Ext(x)_z : x \in z\}$, and to demonstrate that **N** satisfies all of Peano's first order axioms;
2. it is possible to prove the existence of a usual inductive set.

It can be easily seen that there is a problem with the explicit formulation of our intuitive background set theory, i.e., to what extent are both strategies – towers and pits – consistent with each other? For instance, how many of the **ZF(C)** axioms can one use to speak about a non-well-founded *infinite* set without a contradiction? Or, how can one compare both of the above-mentioned strategies within one set theory and how much of **ZF(C)** can one use in it?

The first and the most straightforward possibility is to use a negation of **FA**, i.e. the negation of the axiom of foundation. This axiom is independent of other axioms of **ZFC**. The above leads to another form of the axiom of infinity which postulates the existence of a set containing the infinite series "$\ldots \in x \in m \in u \in n \ldots \in z$" of different sets. For instance:

A.18. $\exists z \neq 0\, \forall x \in z.\{\neg(x \cap z) = 0 \wedge \neg(x \in x) \wedge (x \in m \wedge m \in z \rightarrow x \in z)\}$

The last axiom, as well as **A.14.1**, follows from ¬**FA** and, because **FA** is independent of other axioms of **ZFC**, they are consistent with **ZFC**–**FA**. Moreover, every set can be well-ordered in **ZFC**. Therefore, our infinite set z from **A.18**, is isomorphic to an ordinal. There is no obstacle to identifying this set with the inductive set or **N**. Thus, the sets **A.18**, **A14.1** (perhaps even with the additional conditions such as **Tr** and **Suc**) belong to Cantor's Paradise of **ZFC**–**FA**. However, it is difficult to decide at which place they are located without additional conditions and axioms.

The next possibility is to use a non-well-founded set theory from Aczel (1988). Let us consider **ZFC**$^-$+**AFA**+**A.14.1** ∧ **Tr** ∧ **Suc**, i.e. the system consisting of all the usual axioms of **ZFC** except the axiom of foundation, **FA** (regularity axiom), enriched with a form of the anti-foundation axiom, **AFA**. The properties of such systems are examined in Aczel's book; cf. (Aczel, 1988). **AFA** can have many variants. The two most important versions assure the existence of a decoration (at most one – **AFA**$_2$, or at least one – **AFA**$_1$) of every **apg** (accessible pointed graph). **ZFC**$^-$+**AFA** imposes very strong conditions on the identity of sets because, besides the axiom of extensionality, we also have conditions following on from the properties of the relation of bisimulation.[25] For instance, in the version of **AFA** where there is only one decoration of every **apg**, a graph corresponding to infinite set z from **A.14.1** ∧ **Tr** ∧ **Suc**, seems to be the same as one graph corresponding to the non-well-founded set $\Omega = \{\Omega\}$. Thus, both sets should be identical in this system. However, Ω is a finite set. Therefore, **A.14.1** ∧ **Tr** ∧ **Suc** seems to be inconsistent with this system. The same situation can be found with **A.18.**

25 One and the same – from the point of view of the axiom of extensionality – set can be a decoration of many **apgs**. The sets which have the same **apg**'s structure are identified. The relation of a bisimulation is a kind of isomorphism between **apgs**. The use of only the axiom of extensionality in a non-well-founded set theories cannot decide which sets of the type "Ω" are identical.

However, there are some other possibilities in **ZFC**$^-$+**AFA** to consider infinite non-well-founded sets. In this system sets containing infinitely descending sequences (as well as infinite circular sequences) of the length corresponding to any ordinal number *alpha* from Cantor's Paradise, exist; cf. Nitta *et al.* (2004). Thus, in any case, one can find a graph (different to an Ω-graph) corresponding to our axioms **A14.1** or **A.18**.

One can also use a version of **AFA** called **AFA**$_1$, in which every **apg** can have at least one decoration. However, in this case, every non-well-founded set is only quasi-non-well-founded because it is isomorphic to an ordinal. (Let us recall that we use a theory with the axiom of choice, **AC**, which is equivalent to the statement that every set can be well-ordered.) Therefore, every infinite non-well-founded set belongs to Cantor's Paradise in **ZFC**$^-$+**AFA**.

Obviously, one can resign from using both **AFA** and **FA** in **ZFC**. However, if there is still **AC** and every set can be well-ordered, any other infinite set belongs to Cantor's Paradise. Thus, the only possibility to consider an infinite set which is outside of Cantor's Paradise (i.e. the class of "regular" cardinal numbers) is to use **ZF**–**FA** in which we can construct (an essential part) of Cantor's Paradise. In this last case, the properties of the presented axioms indicate that in some cases it is difficult to judge which sets are equinumerous without additional axioms. Nevertheless, in every case, it is possible to define a minimal infinite set in a usual way, i.e., the given set is minimal iff it is equinumerous with a subset of every set which corresponds to the given axiom or iff it is contained in every such set. This concerns both well-founded and non-well-founded sets.

The next question arises: are all such minimal sets equinumerous? We know that equinumerosity is not a universal or absolute concept which is completely separate from a formal theory and a model of it even in **ZFC**. For instance, the paradox of Skolem informs us that there are denumerable models of **ZFC**. Moreover, we know that the Continuum Hypothesis (**CH**) and General Continuum Hypothesis (**GCH**) are independent sentences from the axioms of **ZFC**, i.e. there are models of **ZFC** which satisfy (**G**)**CH**, as well as some that do not. One can force the existence of even infinitely many infinite sets between, for example, *aleph*$_0$ and c. There are also sentences (axioms) concerning the existence of large cardinals. From these facts, it follows that there is not one Cantor's Paradise, even in the context of **ZFC**.

The last possibility which I am going to consider is a system **ZF** without **AC**, **FA** and the usual axiom of infinity enriched with an axiom like **A.18** $\wedge$ **Tr** $\wedge$ **Suc** or **A.14.1** $\wedge$ **Tr** $\wedge$ **Suc**. In this system, one can try to construct a non-standard "Cantor's Paradise". For instance, as above, we can define *initial segments* of non-well-founded sets. Therefore, one can develop a theory of infinite sets and create Cantor's Paradise without the usual axioms of infinity. However, **A14.1** without **TR** and **Suc** has indefinite cardinality.

We say that "a non-well-founded set can be well-founded in the given formal system **S**" iff there is a one-to-one correspondence between this set and a well-founded set in **S**. There are two kinds of theories possible. The first kind of theories contains only non-well-founded infinite sets which can be well-founded. The second – contains a set which cannot be well-founded. Such a set has an undefined infinity and can be placed ("forced") anywhere in Cantor's Paradise.

5 God's perspective: infinite sets and objects

There are (infinitely) many infinite mathematical objects. The infinity of such objects as "Euclidean surface" in Tarski's elementary geometry, real numbers or straight line is seen from the outside (i.e., from the point of view of a model) of many formal theories because the objects are infinite, as some formal theories do not have finite models and "in" the given theory there is no axiom of infinity. We can operate with such objects without any problems because we used to see them from God's point of view: we do something which is not possible for us – finite, limited and mortal human beings.

From the above considerations, it follows that God's point of view in modern set theory relies on the Platonic attitude taken by working mathematicians which consists of many Platonic methods. To the essence of such methods belongs the treatment and unavoidable acceptance of mathematical objects: finite as well as infinite. For instance, even the use of classical logic belongs to the methods, as argued L. E. J. Brouwer. The change of logic cannot remove other Platonic methods from mathematics. For instance, in categorial set theory with intuitionistic logic, there are still many other Platonic attitudes present.[26]

On the other hand, there is no way to construct infinite objects with the use of only finite sets or other finite objects. However, there is a possibility to prove the existence of an infinite object, assuming the existence only of a singleton; Król (2006), Chapter 9.3 *The One and the Dyad in Plato's 'Parmenides'*.

---

26 More information with the relevant argumentation, can be found in Król (2006). I do not consider, in the present work, the axioms of infinity and natural numbers objects in categorial set theories. Categorial set theories should be classified in the scopes of the present section as "infinity seen from the outside". There is also a strict dependence (i.e., an isomorphism) between the traditional theoretical set models of **ZFC** and the models of categorial versions of this theory.

We can try to grasp many ancient convictions concerning some infinite objects, e.g., the (potentially) infinite divisibility of a continuum, using formal tools. For instance, one has:

A.2.1. $\exists c \forall y \subsetneq c . y \neq 0 \wedge \exists x \neq 0 . x \subsetneq y.$

Thus c is infinite when it is well-founded. This indicates the next possible axioms of infinity in set theory:

A.2.2. $\exists c \forall y \subsetneq c . y \neq 0 \wedge \exists x \neq 0 . x \subsetneq y \wedge x \in c;$
A.2.3. $\exists c \forall y \in c . \exists x\, (x \in y \wedge x \in c).$

From the point of view of theology, in which actual infinity is reserved solely for God, infinite mathematical objects seem to be some fictions or shadows of the ideas in God's mind. One more possibility is the divinization of infinite objects. The last process is exemplified by the so-called *divinization of space*.[27] Koyré describes the views of Joseph Raphson, Newton and others. For instance, he writes:

> Raphson is by no means Spinozist. On the contrary, More's distinction between the infinite, immovable, immaterial extension and the material, mobile and therefore finite one is, according to him, the sole and only means of avoiding the Spinozistic identification of God with the world. (Koyré (1957), p. 191. Cf. also Raphson (1702))

Raphson sees that infinite space has the same properties as are usually attributed to God: indivisibility, actual infinity, absolute immovability, pure actuality, space (and God) are all-containing, all-penetrating, incorporeal, immutable, one in itself (i.e. simple), eternal, the most perfect, incomprehensible to human beings, space is an attribute of the First Cause; cf. Raphson (1702), pp. 194–200. In the same way, infinite absolute space is *sensorium Dei* for Newton.

However, it is an obvious fact that human beings can "step into the shoes of God" and to consider actually infinite objects which have properties which are not in common with God. There are infinitely many infinite mathematical objects which are not God and actual infinity is not reserved for God as His property simply.

27 Cf. Koyré (1957), Chapter VII.

Bibliography

Aczel, P. (1988), *Non-Well-Founded Sets*, Centre for the Study of Language and Information, Leland Stanford Junior University, USA.

Bernays, P. (1958), *Axiomatic Set Theory. With a historical introduction by A. A. Fraenkel*, Amsterdam: North-Holland Publishing Company.

Clagett, M. (1968), *Nicole Oresme and the medieval geometry of qualities and motions. A treatise on the uniformity and difformity of intensities known as Tractatus de configurationibus qualitatum et motuum*, edited with an introduction, English translation and commentary, by Marshall Clagett, Medison, Milwaukee, and London: The University of Wisconsin Press.

Cross, R. (1998), "Infinity, continuity, and composition: The continuum of Gregory of Rimini", in *Medieval Philosophy and Theology*: 7, 89–110.

Dauben, J. W. (1977), "Georg Cantor and Pope Leo XIII: Mathematics, Theology, and the Infinite", in *Journal of the History of Ideas*: 38(1), 85–108.

Dauben, J. W. (1979), *Georg Cantor: his mathematics and philosophy of the infinite*, Boston: Harvard University Press.

Dauben, J. W. (2004) [1993], "Georg Cantor and the Battle for Transfinite Set Theory", in *Proceedings of the 9th ACMS Conference*, Santa Barbara: Westmont College, CA.

Davenport, A. A. (1997), "The Catholics, the Cathars, and the Concept of Infinity in the Thirteenth Century", in *Isis*: 88(2), 263–295.

Euler, L. (1797), *Introductio in analysin infinitorum, editio nuova*, vol. II, apud Bernuset, Delamolierre, Falc. & Soc.. Retrieved from http://books.google.pl/books?id=hQ4OAAAAQAAJ&printsec=frontcover&hl=pl#v=onepage&q&f=false

Euler, L. (1990), *Introduction to Analysis of the Infinite*, Book II, translated by John D. Blanton, New York, Berlin, Heidelberg, London, Paris, Tokyo, Hong Kong: Springer-Verlag.

Fraenkel, A. A., Bar-Hillel, Y. and Levy, A. (1973), *Foundations of Set Theory*, Amsterdam, London: North-Holland Publishing Company, IInd edition.

Fraenkel, A. and Levy, A. (1976), *Abstract Set Theory*, Amsterdam: North Holland Publishing Company.

Grant, E. (1982), *Much Ado About Nothing. Theories of Space and Vacuum from the Middle Ages to the Scientific Revolution*, Cambridge, London, New York, New Rochelle, Melbourne: Cambridge University Press.

Heath, T. L. (1896), *Apollonius of Perga: Treatise on Conic Sections with Introductions Including an Essay on Earlier History on the Subject*, Cambridge: Cambridge University Press.

Koyré, A. (1957), *From the Closed World to the Infinite Universe*, Baltimore and London: Publications of the Institute of the History of Medicine, The John Hopkins University, Third series: the Hideyo Noguchi Lectures, volume VII.

Król, Z. (2006), *Mathematical Platonism and Hermeneutics*, Warsaw: IFiS PAN.

Król, Z. (2015), *Platonism and the Development of Mathematics. Infinity and Geometry*, Warsaw: IFiS PAN.

Nitta, T., Okada, T. and Tzouvaras, A. (2004) *Classification of non-well-founded sets and an application*, http://users.auth.gr/tzouvara/Texfiles.htm/non-well.pdf.

Neumann, J. von (1923), "Zur Einführung der transfiniten Zahlen", in *Acta Lit. ac So. Univ.... (Szeged), Sectio Sc. Math.*: 1.

Proclus, (1987), *Proclus' Commentary on Plato's Parmenides*, translated by G. E. Morrow and J. M. Dillon, with Introduction by J. M. Dillon, Princeton: Princeton University Press.

Raphson, J. (1702), *On the Real Space or the Infinite Being*.
Robinson, A. (1966), *Non-Standard Analysis*, Amsterdam: Studies in Logic and the Foundations of Mathematics, North-Holland.
Zermelo, E. (1908), "Untersuchungen über die Grundlagen der Mengenlehre. 1", in *Math. Ann.*: 65, 261–281.

Uwe Meixner

No Life without Time

1 Concepts of existence

In order to make the point of this paper, some preparations – explications and distinctions – are necessary. The first concern is the concept of *existence*. In fact, there is more than one such concept. There is a semantical concept of existence, and several ontological ones. The many concepts of existence are not always carefully distinguished.

According to the semantical concept of existence (which is precisely the concept of existence employed in so-called "free logic", but only in non-quantified contexts), "N exists" means as much as "'N' refers to something" (for example, "Pegasus exists" means as much as "'Pegasus' refers to something"). Given the semantical concept of existence, true negative singular existence statements offer no particular problem: "Pegasus does not exist" is true because "Pegasus" does not refer to anything; "The present king of France does not exist" is true because "the present king of France" does not refer to anything. However, the semantical concept of existence cannot be *the* (only) concept of existence. This is so because "exist" is, according to the semantical concept of existence, a predicate of *names* (contrary appearances notwithstanding), not a predicate of non-linguistic objects, and because we usually *do* intend "exist" to be a predicate of non-linguistic objects (that is: we usually do intend it to be a predicate *properly speaking*, a *real* predicate).

The ontological concepts of existence are characterizable without speaking about – without referring in any way to – singular terms. There are two ontological concepts of existence: (1) The (real) predicate "*x* exists" may mean as much as "*x* is [identical to] *something*"; this interpretation of "exist" is used, for example, in the following statement: "Some possible world exists which is not merely possible, and some possible worlds exist which are merely possible." (2) The (real) predicate "*x* exists" may mean as much as "*x* is [identical to something which is] *actual*"; this other interpretation of "exist" is used in the following statement: "Some possible world exists and, consequently, it is not merely possible, and some possible worlds do not exist and, consequently, they are merely possible."

Of the two concepts of existence, the second one (described in (2)) is far more interesting than the first (described in (1)); for according to the first concept, it is a logically necessary truth that everything exists. If it is logically necessary that everything exists, then true negative singular existence statements can only be

https://doi.org/9783110594164-007

had if (a) one allows that names may name nothing – which is contrary to classical logic – and if (b) one switches to the semantical concept of existence at least in singular existence statements – which is an *ad hoc* measure in view of the fact that singular existence statements, whether positive or negative, are, as a rule, intended to have an ontological meaning.

Moreover, the intended ontological meaning of singular existence statements is, usually, not the one which is in accordance with the first concept of existence; rather, it is the one which is in accordance with the second. "I exist", uttered by me, has an ontological meaning; but does "I exist" mean that I am (identical to) *something* (namely, myself)? It does not; for otherwise I, for one, would not be ready to add "but I might not have existed" to "I exist". It is logically impossible that *I – this person – am not* (identical to) *something*; it is only possible that *I am not* (identical to something which is) *actual.* I do not change the meaning of the word "exist" from one sentence to the next, and I do not wish to assert falsehoods. Thus, in saying "I exist, but I might not have existed" I am either saying that I am actual and might not have been actual, or I am saying that I am something and might not have been something. And the first has the considerable advantage of being possibly true.

2 Actuality

Actuality is a close companion of possibility. Actuality logically includes possibility. In fact, actuality can be defined on the basis of *possibility* and *mere possibility* as follows: *x* is actual if, and only if, *x* is possible, but not merely possible. If existence is identified with actuality, then many true negative singular existence statements offer, again, no particular problem: "Being a unicorn does not exist" is true because the property of being a unicorn is not actual, in other words: because no instance of it is actual; "The state of affairs that the sun revolves around the earth does not exist" is true because that state of affairs is not actual, in other words: it does not obtain, it is not a fact. "The assassination of Hitler does not exist" is true because that event is not actual, in other words: because it did never happen. What philosophers often balk at are *nonactual individuals* (in the narrow sense): individuals (narrowly conceived, therefore non-events) which are merely possible or even impossible. Philosophers who reject nonactual individuals do not wish to be "Meinongians". The discussion of Meinongianism is not a concern of this paper. Let me just say that I find nothing particularly objectionable in nonactual (or, for that matter, *nonexistent*)[1] individuals as long as one identifies them with the merely possible individuals;[2] *impossible* individuals (individuals that *cannot* be actual), which *full* Meinongians accept (thus moving beyond possibilism), I do find ontologically problematic and somewhat hard to defend.

I move on to considerations regarding actuality which are more important in the present context than Meinongianism. *Actuality* is one-sidedly entailed by *consciousness* and by *aliveness*; aliveness, in turn, is one-sidedly entailed by consciousness. In other words: Necessarily, nothing is alive or conscious which is not actual, but not vice versa; necessarily, nothing is conscious which is not alive, but not vice versa.[3] Now, the basic assumption of this paper is that there is no such thing as *timeless* consciousness, aliveness, and actuality; actuality, aliveness, and consciousness are *essentially time-related*.

3 Three ways of time-relatedness

There are three ways of time-relatedness: (i) time-relatedness via individual reference, (ii) time-relatedness via quantificational reference, and (iii) indexical time-relatedness. The essential time-relatedness of *aliveness*, for example (the same point could also be made *mutatis mutandis* with respect to *consciousness* and *actuality*), consists in the following: In every interpretation of the predicate "x is alive" which is consistent with its basic sense, the predicate "x is alive" must be understood in one (and only one) of the following three ways:

(i) "x is alive at τ" (where "τ" stands for a non-indexical singular term that refers to a particular time-point or to a particular set of time-points);
(ii) "For some τ: x is alive at τ" (where "τ" stands for a variable that runs over time-points, or non-empty sets of time-points, or both over time-points and non-empty sets of time-points);

1 I note in passing that Meinong identified existence and actuality; this is the identification which I, too, would recommend if one wished to have a predicate of existence at one's disposal which is *not* ambiguous.

2 See my defense of possibilism in Meixner (2006).

3 The following is also true: Necessarily, nothing is actual or possible which is not (identical to) something, *but not vice versa* (the state of affairs that $2+2 = 5$ is something, but it is neither actual nor possible); necessarily, nothing is actual which is not possible, *but not vice versa* (the state of affairs that U.M. is never born is possible, but it is not actual). By putting two shorter chains of one-sided entailment together, we obtain the following longer chain of one-sided entailment (one with a *central link*): x is conscious → x is alive → *x is actual* → x is possible → x is something.

(iii) "x is *now* (*presently, currently*) alive".

The ways of the essential time-relatedness of *aliveness* will be of help in analyzing more complex ways of essential time-relatedness which do not concern *aliveness* properly speaking but are still in the vicinity of it, for example, the time-relatedness expressed by "x was alive": "For some t: t was present before *now* and x is alive at t"; or the time-relatedness expressed by "x is still alive": "x is *now* alive and for some t: t was present before *now* and x is alive at t and for every t' which is such that t was present before t' and t', in turn, was present before *now*: x is alive at t'".

All these remarks can, of course, also be applied to consciousness and actuality.[4] If the essential time-relatedness of actuality, consciousness, aliveness is explicitly specified in one or the other of the three ways pointed out above, then the entailment-chain "x is conscious → x is alive → x is actual" (established in the previous section) needs to be adapted accordingly: (i) "x is conscious at τ → x is alive at τ → x is actual at τ"; (ii) "For some τ: x is conscious → For some τ: x is alive → For some τ: x is actual"; (iii) x is *now* (*presently, currently*) conscious → x is *now* (*presently, currently*) alive → x is *now* (*presently, currently*) actual".

4 The living God

After the preparations in the preceding three sections, I now turn to the main concern of this paper. It is a central teaching of Christianity – in all of its different versions – that God is a "living God". What does that mean? Whatever it means, it is certainly meant to entail the proposition that God is *now* alive (and therefore *now* actual). This is, in fact, what every Christian believes; every single prayer attests to

4 However, here is a reason for doubting that *actuality* is essentially time-related: What about the actuality of *abstract* entities, the actuality of numbers, concepts, propositions? Is not at least *their* actuality *timeless*? The objection is interesting if, and only if, some entities are abstract (which is true if, and only if, *necessarily* some entities are abstract; the truth (if it is a truth) that some entities are abstract is not contingent). Assuming that some entities are abstract, there are two plausible ways to react to the objection: (*A*) One denies that any abstract entity is actual; one asserts that, necessarily, every abstract entity is something (and therefore – qua something – existent), but that, also necessarily, no abstract entity is actual. (*B*) One asserts that, for abstract entities, the predicate "x is actual" – which still means the same as either "x is actual at τ", or "For some τ: x is actual at τ", or "x is *now* (*presently, currently*) actual" – and the predicate "x is (identical to) something" are logically equivalent; in this way, actuality is still essentially time-related, but its time-relatedness has become trivial.

this belief – as well as to the belief that God is *now* conscious and *now* actual (and is not a mere possibility). Now, if there were no Time,[5] it would not be true that God is *now* actual, and not true that God is actual at some time, and not true that God is actual at time τ; and it would not be true that God is *now* alive, or at some time alive, or alive at time τ; and it would not be true that God is *now*, or at some time, or at time τ conscious. The actuality, aliveness, consciousness of God depends on the existence of Time. This dependence is a *sine-qua-non* (or *negative*) dependence; it is also – in view of the essential time-relatedness of actuality, aliveness, and consciousness – an *essential* (or *necessary*) dependence: God cannot – *absolutely* cannot – be actual, alive, or conscious without the existence of Time. Thus, if Time did not exist, then, as a necessary – *absolutely* necessary – consequence, God would neither be conscious, nor alive, nor actual, and one might as well say: he would not exist. If, however, actuality, aliveness, and consciousness are properties which God necessarily has (believers usually take actuality, aliveness, and consciousness to be such properties), then, as a necessary consequence, Time exists just as necessarily as God himself exists necessarily in virtue of his necessarily having those properties.

Obviously, an important question must be answered in order to make the assertions in the previous paragraph fully intelligible: What does it mean *that Time exists*? (If we come to know this, we will, of course, also know what it means *that Time does not exist*.) Minimally, *that Time exists* means that Time is *something* – where, necessarily, Time is (identical to) *something* if and only if Time is the set of (all) time-points. This necessary bi-conditional is no great surprise, because Time just necessarily *is* the set of time-points.[6] Alternatively, and rather less minimally, *that Time exists* means that Time is *actual* – where, necessarily, Time is *actual* if and only if (*a*) Time (the set of time-points) is non-empty and (*b*) every time-point was, is (*now*), or will be *present*.

There can be no doubt that Time exists in the sense of Time being actual, and, at the same time, there can be no doubt that the mere assertion "Time is actual" is still far from providing a full ontological description of Time – the above analysis of its actuality notwithstanding. Such a full description will not be provided in this paper.[7] However, here are two additional details about Time which are of

5 I write the word "time" with a capital "T" wherever it serves as a proper name with honorific character.

6 $N(a = b \supset \exists x(a = x))$ is a logical truth, and if $N(a = b)$ is true, then $N(\exists x(a = x) \supset a = b)$ is also true (as a trivial modal-logical consequence). It follows (by elementary modal logic): if $N(a = b)$ is true, then $N(\exists x(a = x) \equiv a = b)$ is true.

7 More can be found in Meixner (1997) and Meixner (2010).

particular importance for the purposes of this paper: (A) The time-points in Time constitute a strict linear order (an order which is structurally just like the order constituted by the elements in any set of real numbers). (B) Each time-point in Time becomes present – singly – and ceases to be present in a succession inexorably proceeding in one single direction along the linear order of Time. Thus, Time is not only actual in the above-defined sense; it is, moreover, (*i*) always true for all time-points t and t' in Time that t' was, is or will be present before t if and only if t' is before t (that is, *timelessly* before t in the linear order of Time), and (*ii*) always true that every time-point in Time is present only once,[8] and (*iii*) always true that exactly one time-point in Time is present.

5 The main worry, and why one need not worry

The main worry is a *theological* worry. If God depended on Time (in the above-described way), would this not make God ontologically dependent on Creation? The ontological dependence of God on Creation would be a highly heterodox consequence, a consequence which should not be accepted, I believe. But how can this consequence be avoided *without* denying that God depends on Time?

There are, basically, two ways of avoiding God's dependence on Creation while accepting his dependence on Time:

(I) Time itself is not created; it is an uncreated part of Creation.
(II) There is uncreated Time and created time; the latter is a part of Creation, the former is not.

Consider solution (I) to the difficulty: According to it, there is just *one* time. The time on which God is dependent is an uncreated part of his essence and at the same time an uncreated part of Creation. Since Time is uncreated, *God does not depend on Creation by depending on Time* (although Time is indeed a part of Creation – but not a created one). Consider solution (II) to the difficulty: According to it, there are *two* times. The time on which God is dependent is an uncreated part of his essence; the other time is created time, a created part of Creation. God does not depend on created time, he only depends on uncreated Time – which, however, is not a part of Creation; therefore, again, *God does not depend on Creation by depending on Time.*

8 In other words, it is always (that is: was always, is now, and will always be) the case that any time-point t in Time which is present was never present already and will never be present again.

Prima facie it might be thought that both solutions contradict divine simplicity because they both take Time to be a part of God's essence. For rebutting this objection, it is necessary to introduce a further distinction. The essence of God *in the narrow sense* is simple. It is the essence God is identical to, as (for example) Thomas Aquinas taught. The essence of God *in the wide sense* is the essence of God in the narrow sense *plus* whatever proceeds ("flows") *per se* (or *eo ipso*) from God's essence in the narrow sense. The essence of God in the wide sense is *not* identical to God; rather, it is identical to *God and his divine life*. If Time is asserted to be "a part of God's essence", then it is, within theological reason, merely asserted to be a part of God's essence *in the wide sense* (namely, in virtue of being a part of what proceeds *per se* from God's essence in the narrow sense); it is not asserted to be a part of God's essence in the narrow sense. For it could not be a part of God's essence in the narrow sense: the essence of God in the narrow sense – being God himself – has no (proper) parts.

The problem with solution (I) is that many theologically interested philosophers, or philosophically interested theologians, are uncomfortable with uncreated parts of Creation.[9] However, are not numbers and universals uncreated parts of Creation? Was not Christ in his divine nature an uncreated part of Creation? If there are no uncreated parts of Creation, then what good reasons are there for us to assume that there is *anything uncreated* beyond so-called Creation? And what good reasons, then, are there for us to assume *anything* about the nature of *the uncreated* allegedly beyond so-called Creation? A perfectly analogical situation in a quite different area of philosophy may serve to highlight the force of the latter two questions: Epistemologists assumed in the past – and many of them still assume – that in cognition we only deal with *representations*. However, if we only deal with representations, then what good reasons are there for us to assume that there is *anything* beyond the so-called representations: something which they represent? And what good reasons, then, are there for us to assume *anything* about the nature of *the something* allegedly beyond the so-called representations?

The problem with solution (II) is that we certainly seem to be talking only about *one* time, not about *two* times, even when speaking about God. And if there were two times after all, one for God and another for us (at least in this world), *what* would be the relation between the two times? This seems to be a question which is not worth the effort of trying to find a plausible answer to it – because it seems unavoidable that the effort is spent in vain. This, if true, would reflect rather negatively on solution (II); it would be a serious drawback to it. However,

9 In Christianity, this is mainly true of the western tradition. The eastern tradition is rather more accepting of uncreated parts of Creation: see Bradshaw (2004), 207–220, 232–238.

since there is a causal relation between God and Creation, one possible answer to the posed, supposedly "embarrassing" question is the following: The two times are related like *game-time* and *player-time*. The state S' of a game (for example, of a chess-game) follows *in game-time* the state S of the game if and only if the action that produces S' follows *in player-time* the action that produces S. Let this bi-conditional define the relationship between game-time and player-time; for the purposes of this paper, no further assumptions are necessary. Note that a large amount of player-time may pass between the two actions which produce two states of the game that are immediately consecutive to each other in game-time (every chess-player knows this). It is also conceivable that a lot of game-time passes between two game-states, with many and various *purely rule-determined* (in other words, purely *law*-determined) game-states in between the two, while the two actions which produce "truly significant change" at the beginning and at the end of the process – because they produce the two game-states in question – are immediately consecutive to each other in player-time.

Obviously, created time corresponds to game-time, uncreated Time to player-time. The idea is certainly not implausible and not without merits.[10] What inclines me nevertheless to reject solution (II) and to accept solution (I) – which, to repeat, is based on the assumption, on the idea that there is only one time, at once an uncreated part of God's essence and an uncreated part of Creation – is a fact of *lived religion*, a fact which neither philosophy nor theology can sidestep: When believers speak of *the living God*, they certainly mean that God is now living *in their time*. In fact, that God is now living *in our time* appears to be one of the messages of God's self-definition in *Exodus* 3, 14: God tells Moses that he, God, is "the I-am". By this, he did not mean to say "I am the *I-am-like-the-natural-numbers-are*". And he did not mean to say "I am *the I-am-in-my-own-sweet-time*". He meant to say "I am *the-I-am-now-in-your-time-and-in-mine*".

If Time is at once an uncreated part of God's essence and an uncreated part of Creation, then we – created beings – partake in every moment of our conscious existence ontologically and cognitively of God's uncreated essence. We should not forget this.

10 It is treated in detail in Meixner (2010).

Bibliography

Bradshaw, D. (2004), *Aristotle East and West. Metaphysics and the Division of Christendom*, Cambridge: Cambridge University Press.
Meixner, Uwe (1997), The Objectivity of Time-Flux and the Direction of Time, In: J. Faye *et al.* (eds.), *Perspectives on Time*, 107–117. Dordrecht: Kluwer.
Meixner, Uwe (2006), *The Theory of Ontic Modalities*, Heusenstamm bei Frankfurt a. M.: Ontos.
Meixner, Uwe (2010), *Modelling Metaphysics. The Metaphysics of a Model*, Heusenstamm bei Frankfurt a. M.: Ontos.

Ludwig Neidhart

God and Time. A defense of God's timelessness

1 Introduction

Concerning the relation between God and time there are two main views, for which I use the names "eternalism" and "temporalism".

Eternalism, as the term is used here, refers to the classical doctrine that God is timeless and outside the timeline in which we live. Eternalism in this strict theological sense has to be distinguished from Eternalism in the philosophy of time, where it is opposed to Presentism.

Temporalism has two main branches: The most radical kind of Temporalism is *Process Theology*, holding that God is not only moving through time horizontally, but so to speak also vertically: God permanently increases His perfection. *Open Theism* is less radical, saying that God remains at all times at the highest level of perfection, moving only horizontally through time. The main tenet of Open Theism is that God lacks complete foreknowledge of the future, so in order to know the outcome of free future decisions He has to wait, and therefore, He is in time just like us. Still more moderate than Open Theism is William Craig's so-called Middle position between Eternalism and Temporalism, which asserts that God was timeless prior to creation but since creation He is in time. Contrary to Open theism Craig also believes that God has complete foreknowledge of the future, but in spite of this He is in time, because He knows at every moment which events are present, and thereby experiences different states of consciousness successively. One could perhaps argue that the ability to undergo change is a typical feature of a temporal being, and so God should be called simply temporal in Craig's concept. If this is right, Craig should be counted as a temporalist, albeit as one of the most moderate members of that group.

We can illustrate Temporalism by a straight line that symbolizes the timeline, and a point moving along the timeline that represents the current position of God in time. In the corresponding illustration of Eternalism we have to put the point that represents God besides the line that represents time, and instead of a straight line we should use the line of a half circle, whose midpoint is the position of the point representing God. Then the point representing God has the same distance to every point on the timeline, which symbolizes aptly the fact that if God lives in timeless eternity He can access with equal ease each point on the timeline by His perceptions and actions.

https://doi.org/9783110594164-008

2 The concept of God

In order to settle the question whether God is inside or outside time, we need as a starting point an appropriate concept of God. A suitable concept seems to be the concept of God in so-called Perfect Being Theology,[1] according to which God is the most perfect being conceivable. This conception is accepted by most theists, including many temporalists. The patron saint of this view is St. Anselm of Canterbury, whose famous definition of God reads: God is *a being than which nothing greater can be conceived*.[2] It seems that this concept can also be connected with the biblical revelation of the name of God, *Yahweh*,[3] which means "he is" in the simplest rendering of the word. This might be an indication that God is pure and unrestricted being, the greatest and most perfect of all beings. Greatness in this definition means of course ontological greatness, which can be illustrated by something's height in the hierarchy of the *levels of being* or *ranks of reality*. The lowest rank is occupied by the so-called *impossible entities* that can't exist in the proper sense, such as square circles or a piece of wooden iron. The subsequent rank is that of *mere possible entities* such as flying horses or golden mountains, that do not exist in actuality but could have been actualized. Above this level follows the rank of *contingent actual entities*, that do actually exist but could also have failed to have actual existence. These entities can be ordered by increasing independence, beginning with the accidents, after which follows the well-known series of minerals, plants, animals and humans, over which the theologians pose the angels. On the highest conceivable rank of this hierarchy one has to put God as a *necessary entity*, for necessity is the extreme contrary of impossibility, which is the characteristic mark of the lowest rank. Now, it seems difficult to reconcile necessity with temporality, because all clear examples of necessary beings known to us (such as Platonic ideas or eternal truths) are timeless. Therefore, the necessity of God, which follows from the definition, could be a first hint that He probably is outside of time.

The just considered levels of reality invite us to take a short look on the so-called ontological argument for God's existence, which I want to present in a most compelling form essentially based on ideas of Leibniz.[4] For every entity, or rather for every *idea* of an entity, there seem to be only three possibilities or options:

---

1 Cf. Rogers (2000).

2 Anselm, *Proslogion* 2: "quo maius cogitari nequit".

3 Cf. *Exodus*, 3, 13–15.

4 Cf. Leibniz (1840), and also Neidhart (2008), pp. 763–766.

1. the idea is *necessary*, which means that the entity exists in every possible world,
2. or the idea is *contingent*, that means: the entity exists in some but not in all possible worlds,
3. or finally the idea is *impossible*, which means that the entity exists in no possible world.

Now, for the classical conception of God, given that God is necessary, the second option must be discarded by definition. So only the first option (according to which God is necessary just as the definition says He is) or the third option (which would be the right one if it turns out that the definition is contradictory) remain. Leibniz has expressed this insight by his famous assertion: "if God is possible", that is: if the third option is also wrong, "then He actually exists", because then the first option must be the right one.[5] Moreover, according to Leibniz, we have a reason to discard the third option, too: impossible ideas always involve some contradiction, but in the idea of God, being absolutely perfect, there are only perfections, which can be described as pure positive and absolute simple qualities, and which, therefore, cannot contradict each other.[6] Thus, in Leibniz's view, the concept of God seems to be an example of a non-contradictory idea, or even the most outstanding example of a consistent idea. If this view is correct, then only the first option remains: God is in fact necessary and so has to exist.

Of course, one can make several objections to this proof, and the same holds true for the other so-called proofs of God's existence: they do not convince everyone. It is not my aim here to discuss at length the pros and cons of the ontological argument and of the other arguments for God's existence. In any case, it seems that all classical arguments for God's existence are very interesting ways of thought, and they might at least fortify the conviction that our belief in God's existence is a reasonable one. What I want to point out is that almost all famous proponents of the classical proofs for God's existence have been deeply convinced eternalists.[7] This, I think, is no coincidence, because the Divine properties needed

5 Cf. Leibniz (1840), p. 177: "si l'être nécessaire est possible, il existe".

6 Leibniz argues this way in Leibniz (1981/2006). See especially Leibniz's argumentation on the last page (p. 577), following the subtitle *Quod ens perfectissimum existit*.

7 For example, Anselm, Descartes, Leibniz and Gödel, the most famous proponents of the ontological argument, have all been staunch eternalists. The same holds for Thomas Aquinas, being the most famous Christian proponent of the cosmological argument. Of course, there are some exceptions: most notably, Charles Hartshorne and William Craig, two famous temporalists, have also defended the ontological and cosmological argument, respectively. But it seems that the arguments proposed by temporalists are not as strong, ambitious and far-reaching as the classical arguments. For example, Craig's version of the cosmological argument leads only to the result that the universe has a cause at its beginning and that this cause was a personal creator, but Craig admits that the proof leaves it open, "whether this creator is omnicient, good, perfect,

in far-reaching arguments for God's existence (such as necessity and simplicity) cannot easily be adopted into a consistent temporalist philosophy, or so it seems to me. So, my point is this: if it would turn out that the existence of a temporal God cannot be rationally defended with equal strength and plausibility as it seems possible to defend the existence of an eternal God, this would be bad news for the temporalists.

However, let us resume the analysis of the concept of God. In order to prove that God has a certain attribute, one only has to show that this attribute expresses absolute perfection. Here I must address the objection that individual judgments concerning perfection are arbitrary. In reply, there seems to be at least some undisputed ontological intuitions about perfection. Consider, for example, the following sequences of adjectives:

- impossible – contingent – necessary,
- perishable – imperishable,
- lifeless – alive,
- unconscious – conscious – self-conscious (which in some systems of philosophy amounts to personality).

I think most would agree that in each row the last adjective describes unre-stricted perfection, which we have therefore to predicate of God. Consider three further examples:

- ignorant (stupid/blind) – knowing (wise/sighted) – omniscient,
- impotent (incompetent, unable) – potent (competent, able) – omnipotent,
- malevolent (bad, perfection-restraining) – benevolent (good, perfection-promoting) – omnibenevolent.

Here also, it seems that we have to predicate of God in each case the last property, and so we get the so-called main properties of God: omniscience, omnipotence and omnibenevolence. It is important to note that Divine omnipotence seems to imply that there is at most one God. For if we suppose that there are two independent acting Gods, both omnipotent, then we obviously get a contradiction because each God should be able to overpower the other. So, this is impossible, unless we

and so forth“ (Craig (1979), p. 152). Likewise, Hartshorne's version of the ontological argument is somewhat deficient in comparison to classical versions, for it accepts only the first step of the Leibnizian version of the argument without reservation, namely that God is either necessary or impossible. But the postulate that God is logically possible, Hartshorne holds, is "the hardest to justify" (Hartshorne (1962), p. 52), he explicitly states that the argument "does not suffice" to exclude impossibility (ibid., 58) and then he adds: "here the other theistic arguments may help" (ibid.). Thus, Hartshorne's ontological proof seems to be incomplete.

say that the two Gods are not acting independently, but are in some mysterious way naturally united, disposing over one and the same supreme power source. But then it seems to be more appropriate to speak about one and the same God occurring in different persons, similar to the Christian doctrine of the Trinity.

Next we need to reflect upon the so-called *simplicity* of God. There are some notions of simplicity in classical theology in which all properties of God are identical with the essence of God and with each other. Such a radical notion of simplicity might be too strong, and anyway, for our purpose we don't need sim-plicity in this sense. What we need is the statement that there might be a variety (probably even an infinity) of different perfections that come together in God and concentrate themselves in a point-like, non-extended Divine essence.

To see that simplicity in this sense is a perfection, suppose that you have a perfectly equipped office, where you can do whatever you might want to do in an office, but for each task you want to do, you have a different device: for writing, a typewriter; for calculation, a pocket calculator; for knowing the time, a clock; and so on. But wouldn't it be more perfect it you had only one single device with which you can do everything? Of course it would, and the reason seems to be that although the office with all these different devices might be perfect considered as a whole, it is not perfect considering its parts: each part, that is each device, isn't most perfect because it is limited in its abilities. So, if an entity has parts, then in order to be most perfect the parts must also be most perfect. But then a multitude of parts seems to be superfluous, for each part would already have all conceivable perfections. Therefore, the most perfect entity should be a simple, unextended entity.

Another consideration leading to the same result is the following. In order to increase a perfection in technology we proceed in two directions. In the first place we try to extend the power and abilities of the device, but secondly there is also the well-known process of miniaturization. We try to concentrate the highest power in a space as small as possible. The reason for this seems to be that an entity being small and having great power is not only more practical, but also more admirable than an entity being big and having equal great power. Therefore, again, the most perfect thing conceivable seems to be a *point-like entity having infinitely great power*. A corollary of this result is, by the way, that *God cannot be a corporal body*, because a corporal body cannot be unextended.

On the other hand, it seems that to be *omnipresent everywhere in space and time* is also obviously a perfection. How can this be reconciled with the property of being unextended? To this one can reply that for a non-corporal (and hence spiritual) entity, presence can be suitably defined in terms of cognition and action. A spiritual entity is *cognitively present* at some point of spacetime, if it can immediately perceive this point; and it is *causally present* at the same point if it can immediately act there. Therefore, the statement that God is omnipresent means simply that He can perceive and act upon everything, and this can be inferred straightforwardly from His omniscience and omnipotence. The overall picture of God we should now have in mind is that God resembles a point, from which different rays come out and connect God with all points in the universe, symbolizing God's perceptions and actions at all locations. This applies of course not only to space but to time as well. Although this picture might suggest already that God is outside space and time, this is not yet made clear by the preceding considerations, because one could suggest that God might be a point *within* spacetime. However, we shall see in the following section why this is not possible.

3 God's timeless eternity

What exactly is eternity? The classical definition from Boethius reads: Eternity is a "simultaneous and perfect possession of illimitable life".[8] It contains three marks, expressed by three adjectives. The first adjective *illimitable* postulates that God's existence extends through every point of time. The second one, *simultaneous*, indicates that all expressions of "life", that is all acting and peceiving, is performed all at once without change and succession. The third adjective is *perfect*. As it stands here, it refers to "possession", and this does not make great sense to me. Therefore, I propose a small change to the classical definition: I would like to shift the adjective such that it refers to the word "life". My definition of eternity then is this: Eternity is a "simultaneous possession of illimitable and perfect life". We shall see in a moment why this alteration of the word "perfect" makes an important difference.

First, note that God matches this definition. *Illimitability* is just the temporal aspect of (and therefore is entailed by) Divine omnipresence. *Simultaneity* of the Divine life means that God acts without interruption in the same manner. This follows from God's immutability, which seems to be just the temporal aspect of Divine simplicity. Finally, the requirement that the life of an eternal entity should be *perfect* means that its ability to act and to perceive should be perfect, ranging over all entities; this of course has to be attributed to God in virtue of His omnipotence and omniscience.

Now we can ask which of the three parts of the definition of eternity (if any) urges us to push God out of the timeline. First, being illimitable is clearly possible

8 Boethius (524/1984), 5, 6, 4: "interminabilis vitae tota simul et perfecta possesio".

for a temporal being; just consider the possibility that a stone lays around for eternity. But also, it seems possible to add the concept of simultaneity to illimitability. The previously mentioned stone, whose "life" consists in lying around, doesn't ever alter its operations and thus "lives its life all at once". To be more precise, we should attribute the property of "simultaneous living its life at once" to every entity that doesn't undergo a succession of different *internal* states. This does not exclude a temporal succession of relations between the entity and the outside world, if this succession is *caused by changes of the outer world alone*. Consider a sun that does not move and shines always in the same way from throughout eternity, while a planet rotates around it. Then any change of relation between the sun and the planet is caused by the movements of the planet alone, and one could say, therefore, despite the changing external relations the sun lives its life simultaneously all at once.

So, if an entity could be temporal despite having an illimitable, simultaneous life, it could only be the addition of the attribute of perfection that pushes the entity outside the timeline. Indeed, one can argue that this is the case. If a temporal entity has an illimitable simultaneous life, which is also perfect to the highest possible degree, the entity must be simultaneously able (a) to act upon and (b) to perceive every event in time. But for a temporal being it seems to be possible only to affect the future and to perceive the past. Therefore, the entity would have to be at the beginning of time in order to be able to act upon the whole timeline exerting influence on every event, while it would have to be at the end of time in order to be able to perceive all temporal events. So, it seems that nowhere on the timeline is it possible to act upon and simultaneously to perceive everything. If this is right, the only possibility seems to be that an entity satisfying the definition of eternity must be outside of time.

But one could ask: How is it even conceivable for an entity to be outside time and in spite of this to be able to act upon and to perceive the world? This question concerns the relation between God and the world. Concerning this relation in general, there seem to be only three general proposals. The first is that God is a physical part of the universe. The second is that the universe is a physical part of God. Apparently, neither the first nor the second proposal seems to be an acceptable option, because in both cases, God and the universe would be parts of one and the same greater spatio-temporal frame of reference. The only remaining third proposal known to me is that *the world is an idea in the Divine mind*. This would mean: the world is not a physical or substantial part, but a mental part, of God.

But then the relation between God and the world is a special kind of a *parallel-*

universe-relation. By this I mean a kind of separation which is neither spatial nor temporal: one cannot travel from one universe to another parallel universe by moving through space or time. Examples for such universes are two dream worlds, two emulated realities in a computer or two real universes as proposed by some interpretations of quantum mechanics.

One can probably explain this suitably by using the idea of emulated (computer-generated) realities. This idea is illustrated by such films as *Matrix* or *The thirteenth floor*, where computer programmers have created an emulated world, and the people in this programmed world think that their world is the real one; they develop a kind of self-consciousness and act independently of the program, guided by their own free will. Moreover, the programmer, by putting some device on his head can enter the computer pro-gram and act immediately with the people therein. I don't think that in fact hu-mans are able to create such realities, but it seems to me that God can and has done something like that.

Now the relation between two emulated universes is the relation of parallel universes; and (what is more important) also the relation between one emulated universe and the real world of the computer programmer is again a relation of parallel universes, although these two universes are not on the same footing as the universe of the programmer has the privileged mode of reality. In an analogous way I see the relation between God and the universe He has created. But if this is the correct description, then of course God is outside of our spacetime and yet He is its creator and supervisor. Possibly He can also somehow enter somehow our universe as the theological doctrine of Incarnation claims.

So here my main argument ends. Before I proceed to the additional arguments, I would like to remark that one should not confound the notion of parallel universes with the notion of possible worlds mentioned before. The difference is that two parallel universes are supposed to coexist both in a fully actualized state within one and the same possible world, while two possible worlds cannot coexist in a fully actualized state. Here I disagree with David Lewis, who somehow seems to identify the two concepts.[9]

4 Additional arguments

Up to now we have discussed God and His absolute perfection and argued that this seems to force us to put Him outside of time. Now we will look in the opposite

9 Cf. Lewis (1986).

direction, that is, down to time, and will discover the imperfection of temporality; thereby, we will get a second argument to put God outside of time again. I want to show this in a kind of ontological meditation, by explaining a symbolic picture I personally have for the flow of time.

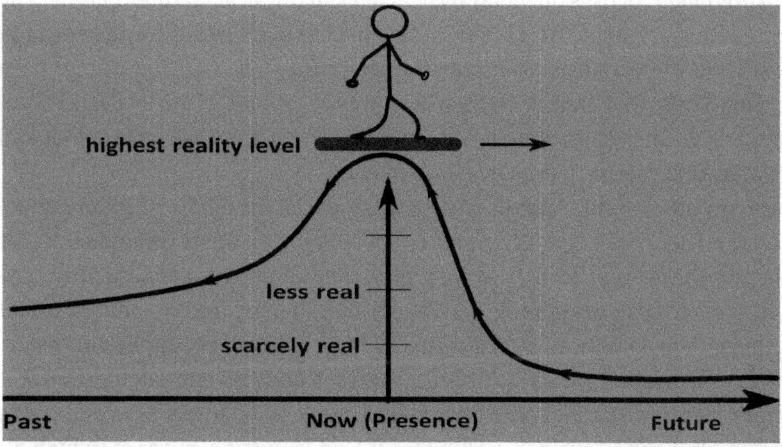

I compare the temporal entity with a surfer on a surfboard surfing on the waves of an ocean, which symbolizes reality. The waters in front of the surfer symbolize the future and the waters behind him stand for the past, while the point of contact with the ocean stands for the present. The height of the water-level symbolizes the intensity or grade of reality, as it appears to the surfer. The level of the distant future is very low, but I emphasize that it is not equal to zero, that is, the future is not totally unreal to us. This is because the future *is doing* something, namely it approaches us, and if it is doing something, it must *be* something real. Then if the future is very near to us (say it is now the tomorrow), it suddenly increases its level of reality, for it throws already its shadow upon us and urges us to prepare for its coming. And if it finally reaches us at the peak of full reality, at the same moment it leaves us and begins to fade down, but it remains forever at a higher level than the distant future, because it remains to us in its effects and also remains visible and explorable to us; that's the historian's job, and history is of course not about nothing but about something real. In short, we have full contact only with the present, less with the past, and even less with the future. And now the question rises: can God be described as such a surfer, with such a restricted contact to reality? Of course not. God must be either the whole ocean of reality, or, if this sounds too pantheistic, He must be the *ground* of the whole ocean, having all the waters of reality completely in His hand.

While the preceding two arguments have been very strong ones in my opinion, the following two kinds of arguments are not so convincing, yet they are important, because there has been widespread discussions about them in the contemporary debate.

The *third argument* focuses on metaphysical or even physical features of time. The most compelling argument of this sort seems to be the following: Time *must* or at least *does* or at the very least *could* have a beginning, whereas God *cannot* have a beginning. Thus, God cannot be bound to time essentially. This is a valid argument, but it is weak because a modest temporalist such as Craig can accept it and even say that God can be in time by His own will.

There are other similar arguments that are based on features of time as revealed by the Theory of Relativity. Some of my fellow eternalists (Brian Leftow for example) use these arguments,[10] while I think they should be avoided. After having spent a lot of time examining the Theory of Relativity and its philosophical implications, I am convinced that the Theory of Relativity only applies to the actual physical universe, and not to spiritual entities and therefore not to God.[11] If for example one argues that according to the Theory of Relativity time and space are inseparably bound together, and hence if God is outside of space (which almost everyone concedes), then by the same token He has to also be outside of time; then one could object that for a spiritual entity this does not apply, for it is perfectly conceivable that a spiritual entity isn't in space although it is in time, because it experiences successive inner states of consciousness. Another argument of this type is that according to the Theory of Relativity there is no universal time, but each of the infinitely many inertial frames has its own time. Then the eternalist could argue that God as creator of the whole universe cannot belong to a particular inertial frame and so cannot be in time. But William Craig has pointed out that there is a possible interpretation of the Theory of Relativity (the so-called Neo-Lorenzian interpretation) according to which there is after all a universal time.[12] I concede that Craig might be right here, but even if he isn't, the Theory of Relativity does not apply to God in any event.

My *fourth and final argument* is the following. If God is outside of time, this would be the best explanation for His foreknowledge described by the Bible. For Biblical

10 Cf. Leftow (2005), p. 67.

11 Cf. chapter 6 of my habilitation thesis *Gott und Zeit* (to appear 2016).

12 Cf. the following writings of Craig: Craig (2010), pp. 163–246; Craig (2001a), pp. 32–66; Craig (2000), pp. 3–126; and Craig (2008, 2001b). For a more detailed treatment of Neo-Lorentian theory including the physical and philosophical side, see Brandes (2010).

evidence consider for example the famous prediction of Jesus, that Peter the next morning will deny him three times before the rooster crows twice.[13] At first glance this seems to be a good argument. But I have to concede that the debate about Biblical arguments is complicated. If one takes the Bible literally there are also Bible verses that seem to support temporalism (for example the so-called repentance of God), so the exegetes have much work to do to reconcile different aspects of the Scriptures, and the same holds for the philosophers, having to deal here with the old question of how to reconcile divine foreknowledge and human freedom. After having spent a lot of time with these questions, too,[14] in the end I think the score here is nearly even, and there is only a slight advantage to eternalism. So, there is much work to be done for only a little reward. On the other hand, after evaluating the remaining arguments, it seems that eternalism is the clear winner. To conclude, I briefly examine the counter-arguments to eternalism, that is, the arguments for God's temporality.

5 Arguments for God's temporality

There seem to be the following major objections to a temporal God:

1. Biblical arguments,
2. arguments based on features of time, and
3. arguments based on properties of God.

As for the *Biblical arguments*, pointing out that the Bible speaks of God as if He is in time, one could reply: The Bible talks anthropomorphically not only as if God is in time, but also as if He has a corporal body and is in space as well, and it's the consensus of almost all theologians that this should not be taken literally. Consider for example Genesis chapter 3, verses 8-9, where it is said that Adam and Eve "heard the voice of the LORD God walking in the garden in the cool of the day: and Adam and his wife hid themselves from the presence of the LORD God amongst the trees of the garden. And the LORD God called unto Adam, and said unto him, Where *are* thou?" Following a literal understanding of this passage God not only doesn't know the future, but furthermore doesn't have complete knowledge of the present either, and moreover He walks around, which implies that He moves through space and has a body.

13 Cf. Marc 14: 30 and 14: 72.

14 Cf. chapter 7 of my habilitation thesis *Gott und Zeit* (to appear 2016).

For the second type of argument, comprising *arguments relying* on the *nature of time*, I have two short examples.

The first is from Wolterstorff: *The eternalist must deny that God exists in any time, but then eternalism seems to be just atheism.*[15]

The second is from Swinburne: A timeless God would exist simultaneously at each point in time, so for example it is one and the same instant at which He knows what I did yesterday, what I am doing today and what I will do tomorrow, and from this it would follow, that all points in time exist simultaneously, which is absurd.[16]

A short reply to Wolterstoff's argument is that it presupposes that "to exist" means "to exist in time", a premise which the eternalist can and will simply deny. Regarding Swinburne's argument, one could reply that Swinburne here invokes the law of transitivity that holds for *temporal simultaneity*, but the simultaneity relation that holds between God and every event of the timeline is of a totally different kind: it resembles (as we saw) the relation that two parallel universes have to each other, although they are not temporally connected. So, Swinburne's argument also fails.

Finally, we arrive at the *most serious arguments for Temporalism*, namely those *based on properties of God*. One invokes here God's *personality*, His *interaction* with temporal creatures and His *omniscience*, and claims that these Divine properties require that He is a temporal being.[17]

Concerning *personality*, it seems that while human personality (which involves memory, anticipation and decision and so on) indeed presupposes temporality, God meets the requirements for personality in a higher, analogous sense: memory and anticipation for example can be replaced in God by His omniscience, which is just a more perfect way of knowing past and future.

Regarding the *interaction* between God and temporal creatures (which seem to presuppose time, especially if God reacts to prayers) one can say in reply that God

15 Cf. Woltersdorff (2001), p. 74: After having stated that the eternalist is committed to assert: "for any time whatsoever, it's not the case that God exists at that time", Wolterstorff comments: "Why isn't that just atheism? ... So I conclude that eternalism is incoherent."

16 Cf. Swinburne (1977), p. 228: "So if the instant at which God knows these things where simultaneous with both yesterday, today and tomorrow, then these days would be simultaneous with each other [...] which is clearly nonsense".

17 For a detailed exposition of these arguments, cf. Craig (2010), pp. 43–55 and 112 (personality), pp. 56–111 (interaction), and pp. 112–133 (omnicience). Craig himself rejects the argument from personality, but empraces the other two.

simultaneously perceives and acts (as we have seen) to the whole of human and cosmic history. Due to His complete knowledge of all free acts in the future, an eternal God also seems to be able to "react" to prayers, which means in some sense that He is able to act "after" having noticed them, provided that the word "after" is used in the non-temporal, mere logical sense, according to which the effect is "after" its cause (or the execution of an action takes place "after" the reason one has for acting), while both can be simultanuous in the temporal sense.

At this point some temporalists make a very interesting move. They assume that God is temporal in the sense that He doesn't know in advance the contingent acts that His free creatures will perform in the future, and they emphasize that this has the advantage that:

1. *it is easier to understand how human freedom can be preserved*, and
2. *it is easier to solve the problem of theodicy*, how it is possible that there is evil in the world in spite of God's omnibenevolence, omniscience and omnipotence.[18]

I agree with the first point. However, the easiest explanation is not always the correct one, and I completely disagree with the second point. Consider a temporal God who doesn't know the future but has comprehensive knowledge of the present, as most temporalists concede. Then suppose God has seen in the 1940s that the Auschwitz concentration camp has been built and that yesterday and the day before hundreds of people have been gassed in the gas chambers. Suppose further that now God sees again a train full of Jewish captives approaching the camp. In this case He must not be omniscient in order to know what is going to happen now, if nothing interferes. So, the temporalist has no less problem here than the eternalist, and I would even say that in fact the temporalist has a much greater problem. The God of eternalism knows in advance the whole extent of evil to come, so He knows also that the evil will not surpass a certain boundary and that at the end the evil will not prevail. Therefore, He can "calm down" knowing that everything will be all right at the end. On the other hand, the temporal God, precisely because He doesn't know all this, has to fear that perhaps all will be totally corrupted and evil may not be compensated. Therefore, He would have a very strong reason to interfere. In short: the more God resembles ourselves, being in time just like us, the more He must also act like we would, and we of course would have interfered in the Auschwitz case. So, it seems that the only promising way to solve the problem of theodicy is the option that God is not as we are, that He transcends our temporally restricted point of view, having other insights from an

18 To the first point, cf. Swinburne (1977), pp. 172–183; for the second, cf. Hasker (2008).

exalted standpoint and, therefore, might see things and have reasons beyond our comprehension. So in the end it turns out that a deeper reflection on the problem of theodicy provides much more support for eternalism than for temporalism.

The last temporalist argument is to invoke God's *omniscience* and to say that God in order to be omniscient has to know how late it is now, and this He can only know if He is in time. To illustrate this argument: A timeless God is like a writer of a theater play, who is absent from the performance. So, while he knows perfectly the sequence of events to come, he doesn't know which act is being played right now, and this seems to be a severe kind of ignorance.

My reply is the following. Temporal (and likewise spatial and personal) indexical words such as *now*, *here* and *I* refer to the spatio-temporal location and to the personal identification of the speaker. Therefore, questions related to these indexicals like *how late is it, where are you, who are you* and so on have to be answered from the perspective of the speaker, anyway. The eternal God could an-swer: *It is now every time* or *it is now eternity*. God's Omniscience refers to His complete knowledge of only *universally accessible facts*. In addition to the knowledge of all or some such facts, every observer has also some knowledge of "indexical facts" that refer essentially to his own perspective: facts that differ for every observer. Therefore, by putting God in time, one would not really increase but only *alter* His knowledge of indexical facts. So, it seems inappropriate to claim that God must be in time to be omniscient.

Yet, the core of the problem is that *God in His divine nature doesn't share with us the same temporal perspective*. This, I think, is not a theoretical problem about God, since it is consistent with our notion of God and His omniscience. But it may be a practical problem for us, because a God that could manage somehow to share with us the same perspective seems to be more attractive for the worshipper than a distant God who is not able to do so.

The solution for this problem in Christian theology is the doctrine of Incarnation. According to this doctrine, God, in addition to His eternal divine nature which remains unchanged, assumes a second created nature and thus comes into time.

This is of course a paradox (although not in the sense of a plain logical contradiction, but in the sense of an unexpected assertion, seemingly difficult to believe or even to comprehend), a paradox that the philosopher and much more the theologian should not try to weaken or trivialize. Such a weakening of the paradox is done a little, I fear, by the temporalists saying that God even in His divine nature is already in time. So, I conclude that eternalism is much better founded in philosophical reasoning and at the end also fits better into Christian theology.

Bibliography

Boethius (524/1984), "Philosophiae Consolatio", edited by L. Bieler, in *Corpus Scriptorum Series Latina*, Vol. 94, Turnhout (Belgium): Brepols.

Brandes, J. (2010), *Spezielle und Allgemeine Relativitätstheorie für Physiker und Philosophen*, Karlsbad: VRI (Verlag relativistischer Interpretationen).

Craig, W. L. (1979), *The Kalam Cosmological Argument*, London: Macmillan Press.

Craig, W. L. (2000), *The Tenseless Theory of Time. A Critical Examination*, Dordrecht / Boston / London: Kluwer Academic Publishers.

Craig, W. L. (2001a), *Time and Eternity*, Wheaton (Illinois): Crossway Books.

Craig, W. L. (2001b), *Time and the Metaphysics of Relativity*, Dordrecht / Boston / London: Kluwer Academic Publishers.

Craig, W. L. (2008), "The metaphysics of Special Relativity: Three Views", in *Einstein, Relativity and Absolute Simultaneity*, edited by W. Craig and Q. Smith, London/New York: Routledge, 11–49.

Craig, W. L. (2010), *God, Time and Eternity*, Dordrecht / Boston / London: Kluwer Academic Publishers.

Hartshorne, Ch. (1962), *The Logic of Perfection*, Lasalle (Illinois): Open Court Publishing Company.

Hasker, W. (2008), *The Triumph of God over Evil. Theodicy for a World of Suffering*, Downers Grove, IL: Inter Varsity Press Academic.

Leftow, B. (2005), "Eternity and Immutability", in *Blackwell Guide to Philosophy of Religion*, edited by W. Mann, Oxford: Blackwell Publishing.

Leibniz, G. W. (1840), "De la demonstration cartesienne de l'existence de Dieu du R. P. Lami (1701)", in *Gottfried Wilhelm Leibniz. Opera Philosophica quae exstant latina gallica germanica omnia*, edited by J. E. Erdmann, Berlin: Eichler Verlag, 177–178.

Leibniz, G. W. (1981/2006), "Ens perfectissimum existit (1676)", in *Leibniz Gottfried Wilhelm, Philosophische Schriften* (1672-1676), Band 3, edited by Leibniz-Forschungsstelle der Universität Münster, Berlin: Akademie-Verlag, 1981. Reprint in Berlin: De Gruyter, 2006, Text No. 80, p. 574–577.

Lewis, D. (1986), *On the Plurality of Worlds*, Oxford: Blackwell Publishing.

Neidhart, L. (2008), *Unendlichkeit im Schnittpunkt von Mathematik und Theologie*, Göttingen: Cuvillier Verlag.

Rogers, K. (2000), *Perfect Being Theology*, Edinburgh: Edinburgh University Press.

Swinburne, R. (1977), *The Coherence of Theism*, Oxford: Oxford University Press.

Woltersdorff, N. (2001), "Response to Paul Helm", in *God and Time: Four Views*, edited by G. Ganssle, Downers Grove, IL: Inter Varsity Press Academic, 69–78.

Peter Øhrstrom
Thoughs on Time, Truth and Transcendence

1 Introduction

A. N. Prior (1914–1969) was the founding father of modern temporal logic and hybrid logic. It is evident that his interest in the relations between time and logic was closely linked to his religious beliefs. As an intellectual and an active member of the Presbyterian community in New Zealand, he often worked with the analysis of the Christian doctrines and ideas in terms of a rational approach to reality. If possible, he wanted to conceive theological systems in terms of the kind of logic that philosophers would normally apply in their attempts to understand reality. In particular, he focused on the tension between the doctrines of human freedom and divine foreknowledge. Prior's analysis of this problem turned out to be extremely fruitful with respect to the development of temporal logic. Using his logical apparatus, he was able to analyse the problem and its possible solutions in a very precise manner. He offered careful and detailed studies of some of the models that make it possible to maintain both the doctrine of human freedom and the doctrine of divine foreknowledge. However, he found that, for various reasons, these solutions have to be rejected (see, Øhrstrom and Hasle (1995, 2006)). Finally, his analysis led him to the conclusion that we have to choose between the two doctrines – at least if the doctrine of divine foreknowledge is understood in the classical way according to which God knows in advance and in all details what we are going to choose in the future.

Prior had a very strong belief in the reality of human freedom. This belief was in fact closely related to his view on time and human existence, according to which there is an important asymmetry between the past and the future. Prior wrote:

> One of the big differences between the past and the future is that once something has become past, it is, as it were, out of our reach – once a thing has happened, nothing we can do can make it not to have happened. But the future is to some extent, even though it is only to a very small extent, something we can make for ourselves. … (Prior (2014d)).

Facing the apparent conflict between the doctrines of human freedom and divine foreknowledge, Prior felt that he had to hold on to human freedom and reject the classical notion of divine foreknowledge. This was probably also his main reason for leaving the Presbyterian Church when the Prior family moved from New Zealand to Manchester in 1958 (see, Kenny (1971), Hasle (2012)). However, even after having left the Presbyterian community and bringing his church activities to

https://doi.org/9783110594164-009

an end, he kept working with theological problems related to temporal logic. In fact, one of his most famous papers on such topics was 'The Formalities of Omniscience', which first appeared in April 1962 (now reprinted with modern formalism in Prior (2003), pp. 39–58). It seems that he held the view that he had stated in a diary note during a much earlier religious crisis in 1942:

> Theology is an illusion, but it is an illusion that is somehow 'close to life', and the study of theological systems illuminates real problems in some way, and that's why I'm interested in it. (Prior (2014a), p. 2)

So even if he did not see himself as a religious believer in the 1960s, Prior could still relate to and reflect on many of the claims made in theology, since they turn out to be 'close to life'.

The aim of this paper is to investigate Prior's argument more closely. It will be demonstrated that the argument can be turned around, which means that it becomes an argument in favour of transcendence in general, and divine foreknowledge in particular.

In section 2, Prior's argument will be considered in its historical context. In section 3, the argument will be presented in terms of Prior's tense-logical formalism. Section 4 offers a further discussion of Prior's argument. In section 5, the argument will be turned around, thereby becoming an argument in favour of some kind of transcendence allowing future contingents to be true now. In section 6, we consider the aspects of transcendence and metaphysics related to the idea of a true future. It will be argued that this view may be presented as a modern version of the view that was defended by William of Ockham (c. 1280–c. 1349) and Luis de Molina (1535–1600).

2 The problem of divine foreknowledge and human freedom

Prior pointed out that 'a whole line of Christian thinkers, running from Augustine (to trace it back no further) through Luther and Calvin and Pascal to Barth and Brunner in our own day, have attacked freewill in the name of religion' (Prior (2014b)). In particular, Prior concentrated on the classical Calvinistic argument against free will put forward by Jonathan Edwards (1703–1758). Based on Prior's analysis (Prior (1967), pp. 113–116) a brief and slightly elaborated outline of this argument from divine foreknowledge to the denial of human freedom can be presented in the following manner, where *E* is any simple event depending on my actions, like my going to the city or having a cup of coffee:

(1) It is true that *E* will happen tomorrow (assumption).
(2) God knew that *E* will happen tomorrow.
(3) It is necessary that God knew that *E* will happen tomorrow.
(4) It is necessary that *E* will happen tomorrow (or, *E* takes place by necessity tomorrow).

The step from (1) to (2) depends on the premise that God knows everything that is true – i.e., the doctrine of divine foreknowledge.

The step from (2) to (3) depends on the premise that, if something has been the case, it is now necessary (understood as 'now unpreventable') that it has been the case. Prior refers to Edwards on this principle in the following manner:

> 'I observed before', he says, 'that in things which are past, their past existence is now necessary ...' this too late for any possibility of alteration in that respect: 'this now impossible, that it should be otherwise than true'. (Prior (1967), p. 114)

This principle is sometimes called the *Diodorean principle*, since it goes back at least to the Megaric logician Diodoros Chronos (ca 340–280 BC), who presented his Master Argument in favour of fatalism (see, Øhrstrom and Hasle (1995), pp. 15 ff.).

The step from (3) to (4) depends on two basic premises. The first, '*E* will happen tomorrow', follows necessarily from 'God knew that *E* will happen tomorrow', which seems to be closely connected to the idea that God cannot be mistaken. The second is a well-known logical principle that Prior also finds in Edwards' writings:

> Those things which are indissolubly connected with other things that are necessary, are themselves necessary. As that proposition whose truth is indissolubly connected with another proposition, which is necessarily true, is itself necessarily true. (Prior (1967), p. 114)

This is in fact a basic principle in modal logic. The claim is that, if *p* necessarily implies *q*, then if *p* is necessary, *q* will also be necessary. This principle can also be traced back to Diodoros Chronos, and it is still widely accepted. It will be hard to find any logician who will accept a model of common-sense reasoning that violates this principle.

Given that we accept the Diodorean principle as well as the modal principle mentioned above, the argument from (1) to (4) shows that, if it is true now that *E* is going to happen tomorrow, then it is also necessary that *E* will happen tomorrow. Furthermore, we may assume that *E* or the lack of *E* (i.e., non-*E*) occurs tomorrow:

(5) Either it is true that *E* will happen tomorrow or it is true that non-*E* will happen tomorrow.

Combining the argument from (1) to (4) and the assumption of (5) gives us:

> (6) Either *E* or non-*E* takes place by necessity tomorrow.

Since *E* is any arbitrary event related to our actions, the argument shows that what happens tomorrow (i.e., what I do) will take place by necessity – no matter whether I perform the action in question or not – i.e., there is no freedom.

One possible way to react to an argument like this is to point out that it is wrong to confront matters of belief with logic and rationality. Prior wrote:

> In our own day Barth and Brunner, while reviving Calvinism of a sort, make it even clearer than former exponents of that creed have done, that what they are expounding is not pure determinism but a quite paradoxical mixture of determinism and freewill. (Prior (2014b), p. 2)

It seems that, whereas Prior during the 1940s was ready to accept this 'paradoxical mixture', it became increasingly difficult for him to defend it during the 1950s. Being a logician, he had to accept that we must take the logical consequences of our metaphysical positions into serious consideration. This means that there is no other respectable way than to deal with the argument and accept its consequences.

Prior emphasised that, in his argument, Edwards is not maintaining that God's foreknowledge is causing us to act in a certain way. According to Edwards, it is rather the other way around, in the sense that our future acts are causing what God knew. Prior explains:

> Edwards further argues, I think with some cogency as well as ingenuity, that if 'God's Foreknowledge is not the cause, but the effect of the existence of the event foreknown, this is so far from shewing that this Foreknowledge does not infer' (i.e. prove) 'that Necessity of the existence of that event, that it rather shews the contrary the more plainly. Because it shews the existence of the event to be so settled and firm, that it is as if it had already been; ... its future existence has already had actual influence and efficiency, and has produced an effect, Prescience: the effect exists already, and as the effect supposes the cause, ... and depends entirely upon it, therefore it is as if the future event, which is the cause, had existed already'. (Prior (1967), p. 114–115)

Clearly, if a future event is as if it had already existed, it certainly appears to be 'now unpreventable'. For this reason Prior finds it very difficult to accept the classical doctrine of divine foreknowledge without immediately also accepting predestination. However, the wish to make a distinction between God's foreknowledge and God's decree is rather common in theological thought. Prior refers to the Westminster Confession (1647), in particular its section II:

> Although God knows whatever may or can come to pass on all supposed conditions; yet hath he not decreed anything because he foresaw it as future, or as that which would come to pass upon such conditions. (Prior (2014c))

It turns out that the price for accepting divine foreknowledge without predestination is that we have to accept the view that what we as humans do now and in the future will influence what God knew in the past. This may be seen as a high price to pay. It is certainly counter-intuitive that anything we can do today can influence what was the case yesterday. However, instead of speaking about causes and influence, we may alternatively speak of implication, in the sense that what we do now has implications for what prophecies were true yesterday. When formulated in this way, the price seems more acceptable. In fact, this is what we have to hold if we want the so-called Ockhamistic solution, which is based on a denial of the Diodorean principle in the sense that our present and future acts will have consequences for what prophecies were true in the past (see, Prior (1967), pp. 122ff.). Prior had studied the works of William of Ockham carefully; in particular, he was interested in Ockham's famous work *Predestination, God's Foreknowledge, and Future Contingents*, which later appeared in translation (William of Ockham (1969)). Prior's Ockhamistic solution may be seen as a modern formalisation of Ockham's original ideas. In his first formulation of the model, Prior included the idea of a true future (Prior (1966)). However, in his most famous book, *Past, Present and Future* (Prior (1967)), he dropped this idea and based the model on the notion of truth at a moment for a chronicle through the branching time system. There can be no doubt that the former version of the model was closer to Ockham's original ideas than the latter.

3 The classical argument in symbolic form

It is useful to state this classical argument regarding divine foreknowledge and human freedom in terms of Prior's tense-logic. We use $F(n)$ as corresponding to 'it is going to be the case in n days that', $P(n)$ as corresponding to 'it was the case n days ago that', D as 'God knows that', and $\Box$ as corresponding to 'it is necessary that ...' The general principles involved in the argument can be stated in the following manner, letting q stand for an arbitrary proposition:

(P1)	$F(n)q \supset \Box DP(m)F(n+m)q$	(Combined tense principle)
(P2)	$P(m)q \supset \Box\Box P(m)q$	(The Diodorean principle)
(P3)	$\Box(DP(m)F(n+m)q \supset F(n)q)$	(God's foreknowledge cannot be mistaken)
(P4)	$\Box(p\Box q) \supset (\Box p \supset \Box q)$	(Basic modal principle)

Using these general principles, we can make the following deduction:

(1) $F(1)q$	(Assumption)
(2) $DP(1)F(2)q$	(From (1) and (P1))
(3) $\Box DP(1)F(2)q$	(From (2) and (P2))
(4) $\Box F(1)q$	(From (3), (P3), and (P4))

In this way, we have proved

$$F(1)q \supset \Box\Box F(1)q,$$

given the general principles (P1)–(P4). Clearly, the similar result holds for $\neg q$:

$$F(1)\neg q \supset \Box\Box F(1)\neg q$$

As above, we now assume the additional principle regarding the understanding of the future:

(P5) $F(n)q \vee F(m)\neg q$ (The future disjunctive principle)

Given the principles (P1) – (P5), this means that we can prove the following thesis from usual propositional logic:

$$\Box F(1)q \vee \Box F(1)\neg q$$

This means that everything tomorrow happens by necessity. $F(1)q$ and $F(1)\neg q$ will not both be possible. One of them is necessary – and will actually be the only possibility in this regard. In consequence, this is a denial of the possibility of free choice.

The advantage of transforming this classical argument into symbolic form is that the rigor of the formalism makes it possible to state more precisely what we can do if we want to hold on to the doctrine of human freedom. It is evident that anyone who wants to maintain that there is a proper freedom of choice will have to deny at least one of the principles (P1)–(P5).

Prior considered various possible solutions, but he mainly concentrated on the Ockhamistic and Peircean models. As mentioned above, the Diodorean principle (P2) should be rejected as generally valid, if we want the Ockhamistic model. If we prefer the Peircean model, we should reject the combined tense principle (P1) and the future disjunctive principle (P5) as generally valid. The main difference between the two models is that, whereas some future contingents are true when evaluated in the Ockhamist model, no future contingent can be true when evaluated in the Peircean model. Prior discussed both models in details in his

Past, Present and Future (Prior (1967), pp. 113ff.), but personally, he preferred the Peircean model.

In a Peircean model, a future contingent, $F(n)q$, can only be true at an instant t if q is true at all instants on all routes (so-called chronicles) n time units later than t. This means that, in this model $F(n)q$ will in fact be equivalent with $\Box F(n)q$. The idea can be illustrated using a branching time diagram:

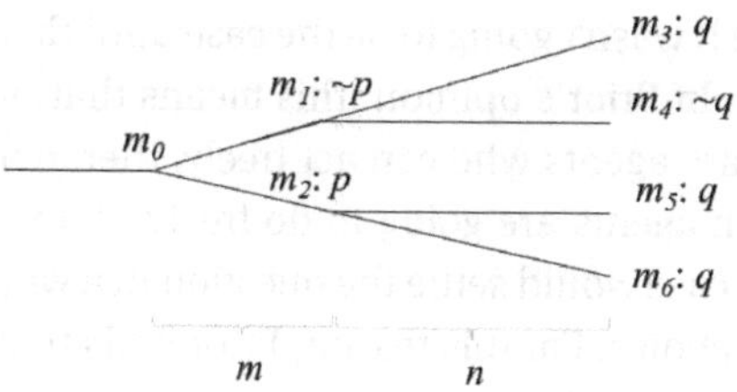

Fig. 1. A branching time diagram

When he was constructing this model, Prior found a great deal of inspiration in C. S. Peirce's writings on the status of statements regarding the future (Prior (1967), pp.128ff). This is why he called his model Peircean. According to this model in Fig. 1, both $\Diamond F(m)p$ and $\Diamond F(m)\neg p$ are true at m_0. (The possibility operator, $\Diamond$, is defined as $\neg\Box\neg$.) However, in this model, neither $F(m)p$ nor $F(m)\neg p$ will be true at m_0. In fact, both propositions are false at m_0. This obviously means that the disjunction $F(m)p \vee F(m)\neg p$ is false as well – a clear denial of the future disjunctive principle (P5). Furthermore, the proposition $P(m)F(m+n)q$ is false at m_2, whereas $F(n)q$ is true at m_2. This means that $F(n)q \supset P(m)F(m+n)q$ is false at m_2, which is a denial of the combined tense principle (P1), since D is redundant in this context (i.e., Dp is true if and only if p is true).

As with the Ockhamist model, there is a price to pay if we choose the Peircean model. This price is first of all that we have to accept the counterintuitive claim that $F(m)p$ and $F(m)\neg p$ in models like the one in Fig. 1 are both false. Furthermore, it is also hard to explain that (P1) and (P5) have to be rejected, since both principles appear to follow from common-sense reasoning. What Prior has to show is that it is reasonable to hold that any statement like $F(n)q$ will be false unless it is necessary – i.e., that no future contingent can be true. In the following, we shall investigate the argument Prior suggested in order to show that this is the case.

4 Prior's argument against true future contingents

In his paper "Contemplation and Action", Prior maintained that, if somebody were able to predict all his future decisions and acts correctly, then his decisions and acts would not be free (Prior (2003), p. 62). Furthermore, he claimed that "if something is the work of a free agent, then it wasn't going to be the case until that agent decided that it was" (Prior (2014d)). In Prior's opinion, this means that, if the world is indeterministic, and if there are agents who can act freely, then not even God can know in advance what such agents are going to do freely. Prior's view is that a divine foreknowledge in this case would settle the question in a way that would destroy the agent's freedom of choice. For this reason, Prior finds that a statement about a future free choice cannot be known as true now. The same can be said about future contingents in general. Prior asks: "... if I now scratch my head, has God always known that I would scratch my head on this occasion?" (Prior (2003), p. 43). According to Prior, there was no truth about this specific scratching before he decided to scratch his head. So even if we assume that God knows every truth there is, it does not follow that God knew about Prior's scratching his head before he decided to do so. This also shows that there is a limited version of the doctrine of divine foreknowledge, which could hold consistently with the doctrine of human freedom. According to this limited version, the doctrine of divine foreknowledge is just the claim that God now knows everything that is true now (which then cannot include any future contingent). It is interesting that this interpretation of the doctrine has been developed further in modern theology as a crucial claim in what has been termed 'open theism' (see, Hasker (1998)).

Prior argued that a similar view can in fact be found in Thomas Aquinas' De Veritate. It appears that Thomas' imaginary objector (and according to Prior, perhaps even Thomas himself) has held "that whatever isn't now-unpreventable hasn't yet come to God's knowledge" (Prior (2003), p. 45).

Prior believed that the world is indeterministic. This means that there are future contingents – i.e., propositions like $F(n)q$ and $F(n)\neg q$ – that are possible but not necessary. However, he pointed out that the truth-values of propositions, like 'Eclipse will win', regarding a future horse race, 'lie on the table until the race is run', and added:

> ... their 'wait and see' character so infects whatever compounds they enter into that the present-tense assertion that such a proposition is now true has itself this 'wait and see' character and must just lie on the table until the verifying event occurs; ...(Prior (2003), p. 123).

Prior held that future contingents are unsettled in the sense that nothing (or nobody) can settle the question concerning whether they will come true or not. This means that there are no present facts to settle the question, and there is no divine foreknowledge that in principle could settle the question. It should be noted that what Prior discussed is different from what was later known as the Truth-Maker Theory. Prior did not suggest that God's knowledge of something would make it true. It is certainly most likely that he would agree with William Lane Craig in holding that something is not true because God knows it, but rather that God knows it because it is true (Craig (2001)). Prior's point was rather that, if we assume that God knows that Eclipse will win, then we have in fact also assumed that the question about the winner of the race has already been settled. He insisted that future contingents like 'Eclipse will win' are unsettled and still 'on the table.' Actually, he even followed the so-called Peircean solution in claiming that future contingents should in fact be regarded as false:

> 'It will be that *p*' is not true until it is in some sense settled that it will be the case, and 'It will be that not *p*' is not true until it is in some sense settled that not-p will be the case. If the matter is not thus settled, both these assertions, i.e. $F(n)p$ and $F(n)\neg p$, are simply false (Prior (1967), p. 129).

Prior's reason for holding that future contingents are not only 'on the table' but also false is that such propositions have not yet be settled, because they depend on verifying events that have not yet occurred. However, as pointed out by Prior, the Ockhamist will see this as very odd and quite unsatisfactory:

> To the Ockhamist, Peircean tense-logic is incomplete; it is simply a fragment of his own system a – fragment in which contingently true predictions are, perversely, inexpressible (Prior (1967), p. 130).

If the persons *A* and *B* are discussing a future horse race, *A* may say 'Eclipse will win the race' and B may say 'Eclipse will not win the race'. They cannot both be right. Only one of them, *A* or *B*, is right. Of course, none of them knows for sure who it is. This is in fact why betting still makes sense.

Prior's argument against the Ockhamist and in favour of the Peircean solution seems to be rather simple:

1. The propositions 'Eclipse will win the race' and 'Eclipse will not win the race' are true now if and only if they have been settled.
2. None of the propositions 'Eclipse will win the race' and 'Eclipse will not win the race' have been settled.
3. Therefore: None of the propositions 'Eclipse will win the race' and 'Eclipse will not win the race' are true now.

5 Prior's argument turned around

If we agree with the Ockhamist that, when *A* and *B* are betting, one of them is right and the other wrong, and if we find that Prior's argument as mentioned above is valid, then we have to deny at least one of the premises in the argument. The first premise may be seen as very fundamental, since it is the claim that something is true if and only if is settled – i.e., it is in principle knowable to someone that this is how things are. In a sense, this may be seen as basic for the understanding of the very concept of truth. For this reason, we may want to hold on to the first premise. In consequence, the second premise should be denied. In this way, Prior's argument has been turned around – i.e., turned into this argument:

(1′) The propositions 'Eclipse will win the race' and 'Eclipse will not win the race' are true now if and only if it is knowable to someone whether or not Eclipse will win the race.
(2′) One of the propositions 'Eclipse will win the race' and 'Eclipse will not win the race' is true now.
(3′) Therefore: One of the propositions 'Eclipse will win the race' and 'Eclipse will not win the race' is knowable to someone.

Given that the propositions 'Eclipse will win the race' and 'Eclipse will not win the race' are proper future contingents, and given that no human being can know any future contingent, it is tempting to turn it into the following argument:

- The propositions 'Eclipse will win the race' and 'Eclipse will not win the race' are true now if and only if it is in principle knowable to someone whether or not Eclipse will win the race.
- One of the propositions 'Eclipse will win the race' and 'Eclipse will not win the race' is true now.
- The propositions 'Eclipse will win the race' and 'Eclipse will not win the race' are future contingents.
- No human being can know any future contingent.
- Therefore: One of the propositions 'Eclipse will win the race' and 'Eclipse will not win the race' is in principle knowable to someone, who is not human.

This may look almost as a proof of God's existence. However, the conclusion that one of the future contingents is in principle knowable to a non-human being does not imply more than the possibility of such a knowing non-human being. On the other hand, this makes it evident that there is a connection between questions regarding time and truth and questions regarding transcendence. And the conclusion of the argument will, of course, fit nicely into a world-view according to which the doctrine of divine foreknowledge is accepted.

6 Molina and a formal representation of the idea of middle knowledge

There can be no doubt that, if we accept the idea of a true future (i.e., that future contingents can be true), we will to some extent also open the door to further discussions regarding transcendence. At least, this seems clear when we look into the philosophical criticism of the idea.

Nuel Belnap and Mitchell have introduced the term 'Thin Red Line' in order to name the theory of the true future (Belnap (2001), Belnap and Green (1994) and Belnap *et al.* (2001)). In their criticism of the theory, they have considered the following example:

> The coin will come up heads. It is possible, though that it will come up tails, and then later it will come up tails again (though at this moment it could come up heads), and then, inevitably, still later it will come up tails yet again. (Belnap and Green (1994), p. 379)

The semantics of an example like this can be discussed in terms of branching time system:

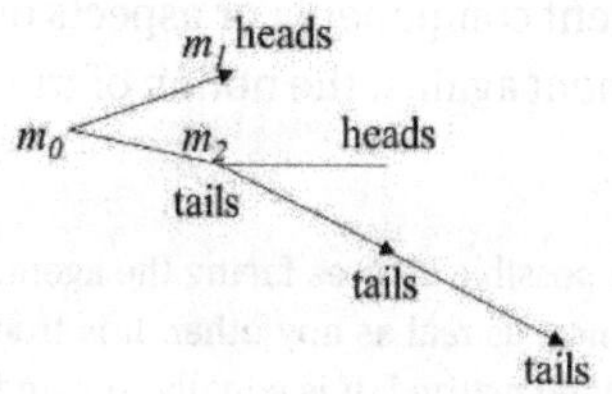

Fig. 2. A branching time diagram corresponding to the example suggested by Belnap and Green (1994).

Belnap and Green want to model the semantics of statements like that using Prior's Ockhamistic model, according to which the evaluation of the statements depends not on the moments of time, but also on the chronicle. In this model, there cannot be truth of a future contingent at a moment, but only truth of the statement at a moment *for a certain chronicle or route* through the branching time system. However, this does not appear to be satisfactory. This example clearly shows that, in natural language reasoning, we may need to refer to what is going to happen as opposed to what can happen and what must happen. It is not enough to be able refer to what will be the case assuming a certain course of events. The claim 'the coin will come up heads' is not a conditional, and in common-sense reasoning, we obviously want to assume that it is either true or false. The arrow at m_0 in Fig. 2 indicates that the claim is true. Belnap and Green have pointed out that "our tendency to believe that there is a Thin Red Line is powerful" (Belnap and Green (1994), p. 366). Nevertheless, they find that the idea should be rejected since it is, in their opinion, based on a mistaken metaphysics. However, this rejection seems to be rather problematic. The fact that there is a powerful tendency to hold that the idea of the true future is meaningful could in itself be said to indicate that the idea deserves a more careful exploration. Belnap and Green have argued:

> The fact, if it is one, that at a given indeterministic moment m there is some history such that it is the one that will occur, is not a state of affairs that supervenes upon what is true of particles, tissues or organisms that exist at m. Those of us who do not postulate a Thin Red Line have no need of such a mysterious realm of fact. (We hope you join us in regarding as spurious a reassurance having the form, "but it's only a logical fact". That's bad logic.) (Belnap and Green (1994). pp. 380–381)

This argument is clearly based on the worlds-view according to which the basic components of reality are material (particles, tissues, organisms). Belnap and Green hold that the opposite view would involve the assumption of "a mysterious realm of fact". However, this judgment does not provide any strong evidence against the view that there are important transcendent components or aspects of reality. Belnap later developed his world-view argument against the notion of true future in the following manner:

> There is no real choice without the reality of alternative possible choices facing the agent. Each of these possibilities is, before the moment of choice, as real as any other. It is true and important that at most one of these possibilities will be realized. It is equally true and equally important that none of these possibilities is a ghostly image of some specially distinguished one among them that some philosopher might label 'the actual choice'. This form of actualism is a bad idea. (Belnap (2001), p. 2)

It is interesting that Belnap finds that 'a ghostly image' would be needed if some future contingents are true now. It seems obvious that his reasons for holding that the idea of a true future is 'a bad idea' are related to his unwillingness to accept the metaphysics that seems to be behind the idea. Clearly, what Belnap has termed 'a ghostly image' of what is going to happen may just be what a believer would express in terms of the doctrine of divine foreknowledge. There is no convincing argument against a true future model. In their analysis, Borghini and Torrengo argued that 'the future is not only as settled as the past; it is also as *contingently settled* as the past' (Borghini and Torrengo (2013), p. 123). Furthermore, they have tried to show that "indeterministic physical laws are best explained when we posit a thin red line" (ibid.).

The metaphysical assumption behind the idea of the thin red line, however, turns out to go far beyond the simple notion of what is going to be. Belnap and his co-workers have convincingly argued that, if we want to accept that a future contingent can be true, then we will also have to say something similar about the counterfactual case ((Belnap (2001), Belnap and Green (1994) and Belnap *et al.* (2001)). The above example in Fig. 2 shows that, if we accept 'the coin will come up heads' as true, we would also be ready to accept the possible truth of statements like 'the coin could have come up tails, and then it would come up tails again'. This means that we will also have to operate with a true future in the counterfactual case. In terms of the diagram, this means that we will have to include arrows not only at m_0 and m_1, but also in the counterfactual case m_2. What we need is in fact a function, TRL, from the set of moments in the branching time diagram and in to the set of chronicles or routes through the diagram (linear subsets). For any moment, m, $TRL(m)$ will be the chronicle through m that corresponds to what would count as the true future at m.

Belnap and his co-workers apparently find the metaphysics of the TRL-theory odd. However, it is interesting the Luis de Molina (1535–1600) defended an idea that is in fact very close to this theory, based on his analysis of the logical relation between the doctrines of divine foreknowledge and human freedom. Molina held that God has so-called middle knowledge. This means that God not only knows what is going to happen, but also what would have been going to happen in any counterfactual situation (Craig (1989)). I have elsewhere argued that Molina's thoughts can give rise to a clear and important response to the criticism of the true future as it has been formulated by Belnap and others (Øhrstrom (2014)).

Molina's argument can be illustrated using a certain passage from the Old Testament in the Bible, I Kings 23: 10–12.[1] In Molina's words:

> David consulted the Lord about whether Saul was going to descend upon Keilah, and the Lord responded, "He will descend". He consulted again, about whether the men of Keilah, who had received nothing but kindness from David, were going to hand him and the men with him over into the hands of Saul. And the Lord responded, "They will hand you over". Notice, God knew these two future contingents, which depended on human choice, and He revealed them to David. Yet they never have existed and never will exist in reality. (see, de Molina (1988))

The branching time analysis of this story is pretty obvious. According to the text, David does not enter the city of Keilah. Yet God knows what the citizens of Keilah

1 I owe this example to Dr. David Jakobsen, Aalborg University.

would choose to do if David had entered Keilah. According to the biblical text, this middle knowledge is even communicated to David:

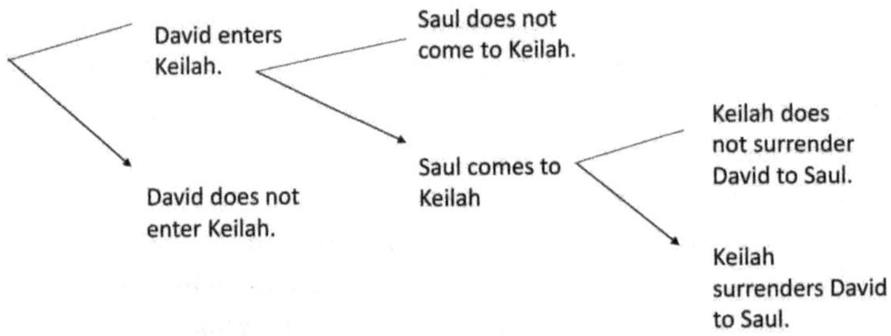

Fig. 3. A Molinistic branching time diagram corresponding to I Kings 23:10–12.

In this way, Molina clearly demonstrated that there is a need for the notion of the true future ('the thin red line') if we want to make a model of the semantics of narratives formulated in natural language, given that we accept the metaphysics on which common-sense argumentation is based (see, Øhrstrom (2014)).

7 Conclusion

A. N. Prior suggested a clear and interesting formal analysis of the logical relationship between the doctrines of human freedom and divine foreknowledge. In fact, this analysis played a very significant role in his development of modern temporal logic. Prior's rejection of the classical version of the doctrine of divine foreknowledge was based on this analysis, and on the view that future contingents cannot be true now since there is nothing to settle such statements. However, Prior apparently also held that a statement is settled if and only if its content is in principle knowable to someone. This means that Prior's argument can be turned around, if we want to assume as one of our premises that future contingents can be true now. By the reversed argument, it then follows that true future contingents can be settled now. Using Prior's line of argumentation, this means that such statements are in principle knowable to someone now. Since no human being could have such knowledge, this would mean opening the door to the possibility of divine foreknowledge and transcendence. As we have seen, Prior held that this would be inconsistent with the idea of human freedom of choice, since a divine foreknowledge of a true statement about the future in his opinion would destroy its contingency and make it necessary. However, this does not have to be so. Contrarily to what Prior maintained, it might be argued to that the future can be contingently settled. This was held by Ockham and Luis de Molina long ago, and it is still a possible and attractive position (see, Borghini and Torrengo (2013), Øhrstrom (2014) and Øhrstrom and Hasle (2015)). In the current philosophical debate, the main arguments against this position are based on a resistance against the metaphysics and transcendence that the assumption of a true future seems to imply (see, Belnap (2001) and Belnap and Green (1994)). However, none of these arguments appear convincing.

Bibliography

Belnap, N. (2001), "Double Time References: Speech-act Reports as Modalities in an Indeterminist Setting", in *Advances in Modal Logic*, edited by F. Wolter *et al.*, Center for the Study of Language and Information, CSLI Publications.

Belnap, N. and Green, M. (1994), "Indeterminism and the Thin Red Line", in *Philosophical Perspectives*: 8, 365–388.

Belnap, N., Perloff, M., and Xu, M., (2001), *Facing the Future. Agents and Choices in Our Indeterminist World*, Oxford: Oxford University Press.

Borghini, A. and Torrengo, G. (2013), "The Metaphysics of the Thin Red Line", in *Around the Tree. Synthese Library, vol. 361*, edited by F. Correia and A. Iacona, Heidelberg: Springer, 105–125.

Craig, W. L. (1989), "Middle knowledge: A Calvinist-Arminian rapprochement?", in *The Grace of God, the Will of Man*, edited by C. H. Pinnock, Grand Rapids, MI: Zondervan Books, 141–164.

Craig, W. L. (2001), "Middle-Knowledge, Truth-Makers, and the Grounding Objection", in *Faith and Philosophy*: 18, 337–352.

Hasker, W. (1998), *God, Time, and Knowledge*, Cornell University Press.

Hasle, P. (2012), "The problem of Predestination: As a Prelude to A.N. Prior's Tense Logic", in *Synthese*: 188(1), 331–347.

Kenny, A. (1971), "Arthur Norman Prior (1914–1969)", in *Proceedings of the British Academy*: LVI, 321–349.

de Molina, L. (1988), *On Divine Foreknowledge*, Ithaca: Cornell University Press.

Øhrstrom, P. (2014), "What William of Ockham and Luis de Molina Would Have Said to Nuel Belnap: A Dscussion of Some Arguments Against 'The Thin Red Line' ", in *Nuel Belnap on Indeterminism and Free Action*, edited by T. Müller, Berlin: Springer-Verlag GmbH.

Øhrstrom, P. and Hasle, P. (1995), *Temporal logic. From Ancient Ideas to Artificial Intelligence*, Dordrecht/Boston/London: Kluwer Academic Publishers.

Øhrstrom, P. and Hasle, P. (2006), "A. N. Prior's Logic", in *The Handbook of the History of Logic*, edited by J. Woods, D. M. Gabbay, Amsterdam: Elsevier, 399–446.

Øhrstrom, P. and Hasle, P. (2015), "Future Contingents", in *The Stanford Encyclopedia of Philosophy*, edited by E. N. Zalta, (2011; updated 2015).

Prior, A. N. (1966), "Postulates for Tense-Logic", in *American Philosophical Quarterly*: 3(2), 153–161.
Prior, A. N. (1967), *Past, Present and Future*, Oxford: Oxford University Press.
Prior, A. N. (2003), *Papers on Time and Tense*, edited by P. Hasle, P. Ohrstrom, T. Braüner and J. Copeland, Oxford: Oxford University Press.
Prior, A. N. (2014a), *Diary entry March 25 1942. The Nachlass of A. N. Prior.* http://nachlass.prior.aau.dk/.
Prior, A. N. (2014b), *Determinism in Philosophy and in Theology. The Nachlass of A.N. Prior.* http://nachlass.prior.aau.dk/.
Prior, A. N. (2014c), *Of God's Plan and Purpose. The Nachlass of A. N. Prior.* http://nachlass.prior.aau.dk/.
Prior, A. N. (2014d), *Some Free Thinking about Time. The Nachlass of A. N. Prior.* http://nachlass.prior.aau.dk/.
William of Ockham, (1969), *Predestination, God's Foreknowledge, and Future Contingents*, translated by M. M. Adams and N. Kretzmann, New York.

Francesco Orilia

The Moral Desirability of Presentism

1 Introduction

In the current debate on time we find a dispute between presentists, according to whom only what is present exist, and non-presentists, who deny that the present exhausts reality; these two groups advertise, we may say, a presentist world and a non-presentist world, respectively. Here I shall put forward a line of reasoning to the effect that a presentist world is morally more valuable than a non-presentist world,[1] and then I shall also speculate on whether this gives us some reason to consider presentism true and non-presentism false, so as to move, as we may put it, from the claim that presentism ought to be true to the claim that it is true.

I have argued elsewhere (Orilia (2016)) in favor of a certain version of presentism, *moderate presentism*, which restricts the claim that whatever exists is present to events and allows for past and future times (understood in a substantialist sense) and for past objects (understood, following Williamson (2002), as "ex-concrete"). My argument will turn around the existence or non-existence of past events and has nothing to do with whether or not there are past (ex-concrete) objects and past times. Hence, strictly speaking, my argument (if successful) supports primarily the desirability of moderate presentism, rather than the standard version of presentism, which we may call *typical presentism*. Similarly, the leap from *ought* to *is* that I have evoked in the previous paragraph regards primarily moderate presentism rather than typical presentism. Nevertheless, for simplicity's sake, the distinction between these two version of presentism can be set aside and I shall thus speak of presentism *tout court* (my proposal can be taken to support presentism in general, even though it supports more directly the moderate rather than the typical version of this doctrine).

The just considered disagreement in the presentist camp is really minor compared to what one finds in turning to non-presentism, for this comes, as is well-known, in various wildly conflicting forms (see, e.g., Markosian (2010)). It includes philosophers who are A-theorists like the presentists, in that they share the belief in an objective present, but are either eternalists or pastists (as both growing blockers and branching futurists may be called). The former acknowl-

1 I have already quickly voiced this thesis in previous works (Orilia (2012, 2014)). In this paper, I try to defend it in detail.

https://doi.org/9783110594164-010

edge in addition to the present both an objective past and an objective future, whereas the latter simply add an objective past. Moreover, most conspicuously, non-presentism includes eternalists of the B-theoretical variety, who acknowledge past, present and future only in a subjective sense, parasitic on objective relations of temporal precedence and simultaneity. For present purposes, however, these divergences in the non-presentist camp are of limited importance and will be ditched as far as possible.

My contention in this paper is inspired by the famous *thank goodness* argument for presentism by Prior (1959, 1996) and it is thus appropriate and perhaps also useful to proceed by first recalling its main point, or at least the one that is relevant here, as well as some well-known anti-presentist reactions to it.

2 Prior's argument and some clarifications

Let us suppose once more, with Prior, that after a severe headache, a demanding exam or a painful visit to the dentist, somebody exclaims with relief: "Thank goodness that's over." From a presentist perspective the unpleasant event no longer exists, whereas it is somehow in the ontological inventory from a non-presentist standpoint. Presentists have thus argued that the relief is appropriate or justified only from their point of view and have appealed to this to back up their ontology. In particular, Prior (1959, 1996) has argued in this way specifically against B-eternalists, whereas Zimmerman (2008), pp. 214–216 and p. 224, n. 5, has raised this issue also against pastists and A-eternalists[2] (beside pointing out that presentists can also justify disappointment for the cessation of pleasure in a way that is not open to eternalists and pastists).

It seems clear that this line can hardly convince non-presentists, for there are ways to see the relief as justified even from an eternalist or a pastist standpoint. The A-eternalist can insist on the objective pastness of the unpleasant experience (Schlesinger (1980)) and similarly can the pastist, whereas the B-eternalist can point out that the experience objectively precedes the belief that one is no longer

2 Although Prior was certainly an A-theorist, some doubts can be raised on whether he was (always) a presentist (see, n. 5 in Hoerl (2015) and references therein). For the purposes of this paper we need not worry about this interpretational matter. For clarity's sake, it is also worth noting that I neglect here a well-known semantic aspect of the argument, which has to do with whether a B-theorist has the resources to correctly interpret "thank goodness that's over" and related expressions. This aspect is perhaps predominant in Prior and in most commentators, but I set it aside, since it is not directly relevant for my concerns in this paper.

undergoing it, whether this belief is taken to be tensed (MacBeath (1983), Mellor (1998), Oaklander (2003)) or tenseless and token-reflexive (Smart (2008)).[3] Moreover, the non-presentist response can be embellished with further sophisticated considerations based on evolutionary theory (MacLaurin and Dyke (2002)) or even statistical mechanics and cosmic thermodynamics (Smart (2008), p. 234).[4]

Before going ahead, it is important to clarify how we should understand tenses or words such as "past" in a cross-theoretical dispute of this sort, in which they may be taken in different ways by different parties. First of all, we should grant that the B-eternalist who does not acknowledge primitive tenses can feel free to read the past tense in her own favorite way. From their perspective, for example, the assertion that Arthur *was* undergoing his most demanding exam, or that the event of Arthur's undergoing his most demanding exam is *past*, can be understood as the assertion that the event of Arthur's undergoing his most demanding exam precedes my tokening these words (or along similar lines). Moreover, we should admit, from a non-presentist perspective, that the present tense can be read in a tenseless way; for instance, from this perspective, the claim that the event of Arthur's undergoing his most demanding exam *is* past (or *precedes* this utterance) can be taken to imply that the event in question is part of reality, that it somehow belongs to the ontological inventory, without however implying that it occurs *now* (simultaneously with my tokening these words), or, so to speak within the present slice of reality (the one containing all the events simultaneous with my tokening these words). I shall take all of this for granted in developing and defending my argument in the following.

3 The argument for the moral desirability of presentism

The anti-presentist reactions to Prior's argument may well convince us that relief makes sense even in a non-presentist world. Yet, they leave a crucial point intact, namely that in such a world all past events exist, are somehow part of reality, so

3 The first option gives rise to the so-called *new B-theory*, whose supporters face the problem of preventing the leap from the existence of tensed propositional attitudes to the existence of tensed facts (see Chen (2011) for a general discussion). The second option avoids this complication and may thus be better for the B-theorist, if the problems typically associated to the other alternative can be overcome (Orilia and Oaklander (2015)).

4 For further anti-presentist rejoinders, see Turri (2013) and Hoerl (2015).

that, *a fortiori*, all past painful events are part of reality; whereas, in a presentist world, no past events are part of reality and hence, *a fortiori*, no past painful events are part of reality.

These considerations, at least *prima facie*, suggest that a presentist world is morally more desirable than a non-presentist world, wherefrom it follows that it is morally desirable that presentism be true and non-presentism false. This is so in the light of a very basic axiological principle, which I believe most us are inclined to take for granted: pain is in itself a negative value, something objectively undesirable in a broadly moral sense; in a nutshell: its absence is morally more valuable than its presence.

Before seeing in more detail how this principle supports presentism, it may be worth considering a couple of objections to it that may come to mind.[5] Somebody might complain that the desirability in question is not really "moral", since pain would be negative even in a world without moral agents, e.g., a world with only lower-level animals, which, though incapable of moral behavior, are nevertheless capable of suffering. It should be conceded however that the negativity of pain grounds a most fundamental moral constraint, namely that one should never inflict unnecessary pain. And thus I think that it is appropriate to qualify as moral the undesirability of pain and the desirability of its absence, despite the possibility of a world with pain and without moral agents. At any rate, for those who insist in considering my use of "moral" out of place, the argument on offer here should still retain some force as an argument for the desirability, *tout court*, of presentism. Next, it could be pointed out that pain can be necessary to attain a positive result, as when we undergo surgery to cure a disease; or that it can be a crucial ingredient of hardships, trials and ordeals whose endurance stimulates new skills and capacities that enhance progress and well-being at both the individual and the social level. Sure enough; the point, however, is that, even if we recognize an *instrumentally* positive role for pain in the achievement of such results, it is not the pain, *qua* pain, that is valuable and desirable, but the results themselves.

Let us then assume that the axiological principle in question is acceptable. Its purchase is not however sufficient to argue for the higher desirability of a presentist world; we need to appeal to the empirical fact that there *was* pain and sorrow, and actually, in doing so, it seems appropriate to bring to the fore not so much relatively minor past offences, such as the headache or the demanding exam considered by Prior, but pain and sorrow of a quite superior magnitude.[6] Unfortunately,

5 Thanks to Tomis Kapitan for having urged them.

6 Ernesto Graziani recommended that this aspect be emphasized.

it is all too easy to find gigantic samples of evil that furnish what is needed here. Think for instance of the Holocaust.

If the Holocaust is real as the non-presentist has it, no matter how much we struggle to make the world a better world, all the intolerable pain, grief and injustice that came with it are still concretely *experienced* in their unbearable fullness somewhere in spacetime. If we seriously concentrate on that, the thought that all that suffering is part of reality should come as a great moral burden, deserving the same empathic sorrow reserved for the suffering around us in our current temporal location. This burden should stand, no matter how relieved we can feel in thinking that the suffering in question lies in a part of reality that precedes ours, and that we can causally contribute to make the stretch of reality ahead of us more comfortable than the one that lies behind. Indeed, the burden may well be compounded by the thought that, precisely because of this lying behind, we can do nothing even in principle to extinguish or at least alleviate the suffering. In contrast, in a presentist world the Holocaust is no part of reality, and thus the empathic sorrow appropriate in a non-presentist world is not called for. There are of course Holocaust memories and documents that upset us and hopefully play a causal role in preventing future evil. And we can and should feel sorry that the Holocaust happened and even be terrified by this thought. But thinking that the Holocaust is part of reality, as in a non-presentist world, is a much deeper burden.

In summary, the *pro-presentist* argument, as we may call it, is this:

(P1) Absence of pain is morally more valuable than presence of pain.
(P2) If there were past painful events, then: (i) they are part of reality, if the world is non-presentist; (ii) they are not, if the world is presentist.[7]
(P3) There were painful events, actually extremely dreadful ones, such as those involved the Holocaust.
Hence,
(C) A presentist world is morally more valuable than a non-presentist world.

In order to resist this conclusion, one can of course challenge the reasons invoked in its favor. In philosophy it is hard to take anything for granted and the controversial area we are investigating is no exception. Thus, if not the empirical fact (P3), (P1) and (P2) can perhaps be contested with some subtle reasoning. Yet, it should be conceded that these two premises are solid or at least that, rather than attack-

7 For clarity's scrupulous sake, let us emphasize this crucial point: it is not being asserted here that, for non-presentists, past painful events are *still* part of reality, for of course they do not claim that such events occur *now*; what is being asserted is that there *are*, tenselessly speaking, such events.

ing them, it is much more interesting to see whether, with the help of (P3), they can really license (C). I see three objections that stand in the way of this result, at least one of which is pretty much obvious and immediate. Let us thus start with it and then move on to consider the other ones.

4 The specular argument objection

We saw that, to reach (C) from (P1) and (P2), (P3) was appealed to. Now, (P3) focuses on the existence of past sorrow. The obvious rejoinder then is an invitation to look at the other side of the coin, namely the existence of past joy, so as to construct a specular *anti-presentist* argument with the opposite conclusion. Here it is.

(P1′) Absence of joy is morally less valuable than presence of joy.
(P2′) If there were past joyful events, then: (i) they are not part of reality, if the world is presentist; (ii) they are, if the world is non-presentist.
(P3′) There were joyful events, e.g., your favorite exciting ones, or even simply the happily routine episodes of ordinary lives conducted in favorable circumstances.
(C′) A non-presentist world is morally more valuable than a presentist world.

Is there a way to choose between the anti- and the pro-presentist arguments? Or should we rather simply admit that they are on a par and thus consider foolish these attempts to establish the moral superiority of one doctrine over the other? If it were possible to calculate the amounts of past (and future) joy and sorrow and discover that sorrow prevails over joy or vice versa, one could perhaps argue that, depending on the outcome, the former or the latter argument should be preferred. But, as far as I can see no such calculation is in view and thus one may suspect that we have a stalemate.

However, it seems to me that, independently of any such calculation, the pro-presentist argument sticks out. For the permanence of suffering that comes with non-presentism seems so intrinsically horrifying that it cannot be compensated by any permanence of joy. To put it otherwise, whatever comfort we may gain from the thought that pleasant past events survive *sub specie aeternitatis*, this can hardly balance the dismay for analogous survival of the unpleasant ones: the dismay prevails, even if in the past there had been overall, let us imagine, more good than evil.

Perhaps this analogy will help us see this. Suppose a powerful demon is about to flip a coin with this in mind: if the outcome is cross, it will bring about that a number of people will enjoy an extraordinarily pleasant experience, but at the same time someone will suffer excruciating pain; if the outcome is head, he will do nothing. Perhaps from a perspective such as hedonistic utilitarianism, one could argue that, if the level of pleasure brought about by the pleasant experiences is sufficiently high, the world is better off with them, despite the simultaneous existence of the pain. But this is a very questionable perspective. It seems to me that most of us would not want an unfortunate fellow in excruciating pain, no matter how high the pleasure of the lucky ones, and thus I imagine you will presumably hope that the outcome is head. Similarly, I think, we should prefer a world in which all past painful events are no part of reality, even if this means that all past pleasant events are similarly erased.

For additional support, we can perhaps also adapt to our case Rawls's (Rawls (1971)) device of imagining an original position wherefrom rational agents are supposed to select principles of social arrangements. Since they are under a veil of ignorance that prevents them from knowing which role in society will befall them, all agents, argues Rawls, would choose an arrangement that grants basic liberties to everybody. This rules out, e.g., a society with a minority of slaves ensuring various benefits to members of the majority, for, no matter how high these benefits would be, nobody would want to run the risk of finding oneself in that minority. Similarly, we can imagine an original position wherefrom we can decide whether or not joyful, but also dreadful, events are preserved in the way non-presentism suggests. Given a veil of ignorance, presumably the non-presentist option would be rejected, since, for all one knows, one could be involved in episodes of excruciating pain or deep sorrow, perhaps in most sections of one's life. The thought that these episodes be parts of reality *sub specie aeternitatis* is too horrifying to license a preference for non-presentism, even though this also grants the preservation *sub specie aeternitatis* of joyful events, or so it seems to me.

May be this further consideration can provide another way to support (C) and thus the pro-presentist argument. In a presentist world, we can conceive that the world as a whole will become *pure*, that is, with no suffering and no evil, yet still with well-being enjoyed by sentient beings. In contrast, this ideal is banned in a non-presentist world, wherein we can at best hope for the purity from some point onward of the temporal slices of reality lying ahead of us. These slices, pure as they may be, however coexist with impure slices full of suffering and evil, such as those containing the Holocaust events. Hence, even from the perspective of this extremely optimistic outlook, a non-presentist world must perforce be, as a whole, an impure world.

But again there is the other side of the coin. Nice as the thought of a pure world could be, one can urge, however, that there are other, less cheerful, but perhaps more realistic, possibilities. One of the options considered by modern cosmology is a "Big Crunch" with which the universe, and all life and sentience with it, comes to an end. This scenario may well be depressing and the non-presentist may point out that her world-view has the advantage of making it more bearable: perhaps there is some comfort in thinking that, even if there is a last slice of the universe, all the preceding slices are part of reality, with life and sentience embedded in some of them. True, not only pleasure and joy, but also pain and sorrow, will be found in these slices. Yet, sad as this may be, one could say, there is certainly more value in this way than in the way brought about in a presentist world, for the Big Crunch scenario in such a world would leave us with nothing at all.[8]

What can the presentist reply? in the first place it should be noted that, in the light of the powerful demon and initial position thought experiments, it is not so obvious that in the Big Crunch scenario non-presentism is more desirable than presentism. But even if it were so, all that this proves is that claim (C) must be slightly weakened, by making it conditional on the falsehood of the Big Crunch hypothesis. We shall go back to this in § 7.

5 The ugly truthmakers objection

Notoriously, presentism has to face a truthmaker problem. It arises once we accept the plausible *truthmaker principle*, according to which (at least some) truths require truhmakers. By its light, true past-tensed propositions, such as the proposition that Caesar crossed the Rubicon, constitute for presentism a difficulty that pastism and eternalism steer clear of. For pastists and eternalists can straightforwardly appeal to past events as truthmakers for them, whereas this move is not open to the presentist. The latter has therefore the burden of extracting from her ontological inventory alternative items that can go proxy for past events in fulfilling this task. (Pastists are in the same boat as presentists when we proceed to consider true future-tensed propositions, such as, let us assume, the proposition that there will be a human expedition to Mars. However, we can ignore this issue here.) One may circumvent this obligation by rejecting the truthmaker principle (Merricks (2007)) or denying that past-tensed propositions are ever really true. But the majority of presentists recognize the obligation (rightly so, in my view) and have presented a rich variety of options in order to meet it. For example, Bigelow's Lucretianism (Bigelow (1996)) appeals to "propositional" properties (e.g., being

8 This worry was raised by an anonymous referee.

such that Caesar crossed the Rubicon), which the world as a whole exemplifies.[9] Keller's heacceitism (Keller (2004)) appeals to present haecceities of past objects and times; Bourne's (Bourne (2006)) and Crisp's (Crisp (2007)) Ersatzism appeal to a precedence relation linking Ersatz times; my own moderate presentism (Orilia (2016)) appeals to the exemplification (possibly by ex-concrete objects) of past-tensed properties of the type *having been F at t*, where *F* is a property and *t* a past time.

Now, and here comes the objection,[10] if items of this sort are admitted, there will also be those that make true, now, past-tensed propositions such as the proposition that there was the Holocaust and more specific Holocaust propositions such as that Ann Frank was murdered. And such "ugly" truthmakers make the world a morally ugly world just as evil past events render ugly the non-presentist world. Suppose for example that Lucretianism is right and accordingly the world has now the property of being such that Ann Frank was murdered. Is such a world any better than a non-presentist world in which the event of the assassination of Ann Frank precedes present events such as my writing these words? According to the objection that we are considering, the answer is negative, for after all, if Lucretianism is right, the ugliness of the world is testified by its exemplifying the propositional property of being such that Ann Frank was murdered no less than by its including, if presentism is wrong, the event of the assassination of Ann Frank. And such a propositional property will never cease to be exemplified, even in a hypothetical ideal future scenario in which there will be no evil, thereby testifying, even *then*, the ugliness of our world. No improvement, the objection continues, is offered by the other solutions that the presentist can offer in responding to the truhmaker problem. If haecceitism is true, Ann Frank's haecceity is and always will be appropriately related to the property of being murdered and to the haecceity of the time of the murder in question, in a way sufficient to make it true that Ann Frank was murdered. And this and other truthmakers of this sort are enough to certify the ugliness of our world even in the ideal scenario. Independently of what the presentist will choose, the objection concludes, analogous considerations will be in play.

Convincing as this line might seem at first glance, it overlooks a crucial point. No matter how close the presentists' truthmakers are to the non-presentists' past events, only the latter involve, so to speak, the "real action." And it is only with the real action that there is, in the unfortunate cases, real suffering. Metaphorically

9 Bigelow draws on Lucretius, who however, as noted by Bigelow himself, takes properties of this sort to be exemplified by sections of matter or portions of space.

10 I owe it to Gregory Landini.

speaking, the truthmakers of true past-tensed propositions, whatever they are, can be compared to films shot when certain events were taking place. When the events are gone, the films remain. If the events were unpleasant, they involved suffering, but no suffering is involved in the corresponding films. If a torturing event is filmed, the film can certainly testify that the victim was in pain, but it can do nothing to keep the victim's pain in existence. Those who watch the film may have an empathic emotional response to it, but this is another matter. This response is not the victim's pain. If presentism is right, fortunately this pain is no longer around. In contrast, if non-presentism wins the day, the pain is with us in the ontological inventory. True, if presentism wins, we still have the film and cannot get rid of it, whereas, given non-presentism, there is no such film, or at least no need to suppose that there is. But certainly it is better to have the film of the torture than the real torture, for only the latter involves the victim's pain. In sum, this objection does not really undermine (C).

6 The radical objection

This last criticism is analogous to the previous one, but it adds a more radical twist to it.[11] The pro-presentist argument, or for that matter the specular anti-presentist argument as well, relies on a hidden assumption, namely that the moral value of a world is based fundamentally on what is real or existent in that world. But perhaps this assumption could be questioned. Perhaps the value of a world depends equally on what exists and on what existed, i.e. on its history, regardless of whether it is understood in a presentist or non-presentist fashion, to the point that there is no difference in terms of value between a presentist world in which the Holocaust existed and no longer exists and a non-presentist world in which the Holocaust is part of reality: both are equally bad insofar as they have the same less than impeccable history, and thus we should have no desire to be in one rather than the other. To put it otherwise: any plausible version of presentism has to be combined with a dynamic view of reality. According to presentism, reality is exhausted by what is present, but, additionally, reality changes. So it is not just that reality contains a truthmaker for the past-tense claim that the Holocaust occurred; that there is this truth means that, sometime in the past, before certain changes brought us to the present time, reality was partly constituted by the Holocaust's

11 The previously mentioned anonymous referee first raised this issue to me by elaborating on the truthmaker objection.

occurring. So, given this, it seems no better for the Holocaust's merely to have occurred (as the presentist claims) than for the Holocaust to be part of reality (as the non-presentist claims). A further, more emotionally loaded, formulation of this kind of concern has been voiced to me by a presentist friend, Tomis Kapitan, as follows: "I tend to be a presentist, but take utterly no comfort in it. I don't think that a presentist world is 'better' in any sense. Even if I agree that past sufferings are no longer real, they *were* real, and that's bad enough to cause me considerable anguish. I don't think that reality would be any worse off if eternalism were true. Similarly, past enjoyments are no longer in existence, and sometimes I get delight in reflecting upon them, but I don't think that the world would be any better if reality is a four-dimensional universe."

This objection is radical, because it undermines at its very foundations the inquiry being pursued here. I do not think, however, that it has a real bite. To be sure, there is a grain of truth in it, but once this element is disentangled from the rest, it should be evident that the objection does not stand. One can and should concede that what happened in the past is relevant for the moral evaluation of a world. Thus, for example, a world in which the Holocaust existed is morally far less desirable than a world in which there was no Holocaust. Yet, in a presentist perspective, the fact that certain events were true cannot have the same relevance in the evaluation as the fact that certain other events are true; the latter have a primacy that the former cannot have. For example, it is bad now and forever that an innocent victim was tortured in a concentration camp and thinking of this can elicit *now* our dismay. Yet, the reality of someone's being tortured, with the excruciating pain of the victim going on *now*, is *worse*. In contrast, in a non-presentist perspective, this can hardly be claimed. To see this, it is useful to focus on the analogy between time and space that is often brought up in clarifying what B-type eternalism amounts to. In the latter perspective, the difference between past, present and future is compared to the subjective distinction between far and near: we are distant in a temporal sense from a torture going on in Auschwitz just as we are far in a spatial sense from, say, a planet in Andromeda; and just as the Andromeda planet is as real as the Moon near us, similarly the past torture in Auschwitz is as real as a present torture going on now. But, if this is so, the past torture is as bad as the present one, just as a torture is bad whether it takes place on the far away Andromeda planet or on the nearby Moon.[12]

12 This is not to say of course that the B-theorist regards spatial distance as the same as temporal distance; the point is simply that for the B-theorist earlier events are real just like spatially distant events. And this point remains no matter how much the B-theorist emphasizes the difference between spatial and temporal relations, e.g. in the way put forward by Oaklander (2015) in

The picture changes a bit in pastism and A-eternalism. For, according to these doctrines, an objective pastness accrues to past events, and this makes them not fully comparable to spatially distant objects. Yet, it does not really matter for our purposes, unless this pastness makes events somehow less real, to the point that past painful events do not involve suffering and past joyful events do not involve pleasure. If so, however, these doctrines become rather uncomfortable, for they almost seem to embrace *contradictiones ex vi terminorum*. Here is how Zimmerman (2008), p. 215, makes this point:

> … if past headaches are to be much better than present ones, these A-theorists [A-eternalists and pastists] must say things like: a headache is only *truly* painful when it is present; yesterday's headache, although it exists, is no longer painful … and that's why it no longer concerns us. … Although this view makes sense of our relief when pain is past … it has less appealing consequences as well. Headaches can exist but not be truly painful.

I have assumed throughout that pastists and A-eternalists do not want to be committed to these past pains that do not hurt. With this assumption they are for present purposes in the same boat as the B-eternalists. From the point of view of all of them, past pain should have the same negative moral weight as present pain. In contrast, in the presentist perspective, one can say that a world in which there was pain is worse than a world in which there was no pain, while admitting that present pain is altogether another matter. Of course, if the pastists and the A-eternalists admit past pains that do not hurt, they *ipso facto* leave, as far as we are concerned here, the B-eternalist's boat and begin to approach the presentist's boat. But they will have to face the concerns raised above.

Despite the above rejoinder, however, it has been suggested to me[13] that an appropriate mental experiment could provide new fuel for the radical objection. Here is how. So far we have compared, so to speak, *our* presentist world and *our* non-presentist world, that is, two worlds both of which have our actual history, a history that unfortunately includes the Holocaust and many other evils. But we may also compare our non-presentist world to a presentist counterfactual world whose history differs from our actual history only because it includes some additional suffering, due, say, to a longer duration of World War II, sufficient for the Nazi to bring to completion their genocidal programs. What is morally preferable, our non-presentist world or the presentist counterfactual world? The supporter of the radical objection may want to suggest here that the former is better, despite the

his recent defense of a Russellian version of the B-theory called “R-theory”, according to which temporal precedence is a primitive unanalyzable external relation.

13 Thanks to Ernesto Graziani for pressing this point.

enormous amount of evil that is erased from reality in the latter alternative. Suppose this answer were correct. This seems to back up the intuition that triggers the radical objection. The additional past evil brought about by a longer World War II, devastating as it might have been, is very small, when compared to all the past evil that is part of reality in our non-presentist world. If, in spite of this, our non-presentist world is better than the counterfactual presentist world, this may be taken to suggest that the moral value of a world depends so much on its history that metaphysical considerations regarding the presentist or non-presentist nature of this history can do little to change the picture.

However, I do not think that this conclusion follows, even if we concede the moral superiority of our non-presentist world over the counterfactual presentist world under consideration. Perhaps, the history of a world matters to a *very large* extent, an extent larger than we might have thought before this thought experiment; and it is certainly worth investigating why this is so, or at least why it seems to us that it is so. In this investigation, we might find out that the minimal axiological principle (P1) is insufficient to back up this role that we assign to history in judging about the moral value of a world. Nevertheless, this does not yet prove that the way in which metaphysically this history should be considered is irrelevant. It remains true that in a non-presentist world a past pain is as real as a present pain in Andromeda and thus (P1), weak as it may be for other purposes, is strong enough to back up the claim that our presentist world is better than our non-presentist world. Perhaps the latter fares better with respect to *other* presentist worlds, but after all we are interested in our presentist world.

7 Can we leap from ought to is?

Suppose that all of this is by and large correct and that we endorse (C). In the light of it we should accept that it is morally desirable that presentism be true. I trust that this result can be considered worthy of note in its own right, but of course its interest increases if it gives us some reason to believe that presentism is true. Does it? Here are my tentative considerations.

The most obvious and immediate reaction to this question, I surmise, is that wishful thinking can hardly settle issues in ontology and that accordingly considerations of the sort I have advanced here are far from proving that presentism is true and non-presentism false. If one comes to accept (C), and yet thinks that from a purely theoretical, ontological, point of view there are compelling arguments against presentism (as many non-presentists hold), the endorsement of (C) may bring some embarrassment, but in itself should not be expected to lead to a conversion to presentism. Consider for example someone who believes that Einstein's relativity theory should be endorsed as part of our best science and that it implies that presentism is false. Such a person presumably will not endorse presentism in the light of (C), even though she might concede that the falsehood of presentism is regretful. However, the matter might be different for those who see the presentism vs. non-presentism dispute as tangled in a theoretical stalemate at the end of the usual round of philosophical arguments in favor or against the options at stake (see, e.g., Sider et al. (2008), chap. 5). Perhaps the pro-presentist argument could incline at least them toward presentism. Given (C), they could come to think, presentism *ought to be* true (provided, one might prudentially add, the Big Crunch hypothesis is not true). And, given the theoretical stalemate, they might thus come to embrace presentism from a *practical* point of view and take it to be actually true. In an analogous fashion, Kant suggests in his second *Critique* that certain propositions that cannot be proven theoretically can be accepted from a practical standpoint.

Moreover, for those who believe in an omnipotent and benevolent Deity or attribute ontological efficacy to values (see, e.g., Leslie (2013) and references therein), the moral superiority of presentism proclaimed by (C) should even be more compelling, perhaps up to the point of leading them to revise their ontology, if it is a non-presentist ontology. From their perspective, it seems to me, the temptation to infer that presentism is true from the proposition that it ought to be true should be particularly high and thus anti-presentist ontological beliefs or sympathies (see, e.g., Leslie (2013), pp. 138-139), if any, should be seriously questioned. Moreover, as they see matters, the ideal of a future pure world with absolute well-being and no sorrow[14] could perhaps be seen as more realistic than the Big Crunch hypothesis; the latter could then be put to rest, unless of course the disappearance of a physical universe is not deemed so important after all, given the faith in a purely spiritual paradise.

8 Conclusion

The argument for the moral superiority of presentism advanced here faces a number of difficult objections. Yet, it seems to me that it withstands scrutiny. At the very least, it offers some interesting food for thought. If its conclusion stands, the

14 At least no sorrow of the innocent, or, following von Balthasar's idea of an empty hell, no sorrow in general.

argument could or should incline toward presentism at least those who think that the presentism vs. non-presentism dispute cannot be resolved on purely theoretical grounds. Moreover, given a fundamental commitment to the ontological efficacy of values, whether theistic or not in nature, it should even more compellingly lead to an endorsement of presentism.[15]

Bibliography

Bigelow, J. (1996), "Presentism and Properties", in *Philosophical Perspectives*: 10, 35–52.
Bourne, C. (2006), *A Future for Presentism*, Oxford: Oxford University Press.
Chen, C. K. (2011), "Subjectivity and the Metaphysics of Time", in *The Possibility of Philosophical Understanding: the Philosophy of Barry Stroud*, edited by J. Bridges, N. Kolodny and W. Wong, Oxford: Oxford University Press.
Crisp, T. (2007), "Presentism and the Grounding Objection", in *Noûs*: 41, 90–109.
Hoerl, C. (2015), "Tense and the Psychology of Belief", in *Topoi*: 34, 217–231.
Keller, S. (2004), "Presentism and Truthmaking", in *Oxford Studies in Metaphysics*, Vol. 1, edited by D. W. Zimmerman, Oxford: Oxford University Press, 83–104.
Leslie, J. (2013), "A Proof of God's Reality", in *The Puzzle of Existence. Why There is Something Rather than Nothing?*, edited by T. Goldschmidt, London: Routledge, 128–143.
MacBeath, M. (1983), "Mellor's Emeritus Headache", in *Ratio*: 25, 81–88.
MacLaurin, J. and Dyke, H. (2002), "'Thank Goodness That's Over.' The Evolutionary Story", in *Ratio n.s.*: 15, 276–292.
Markosian, N. (2010), "Time", in *The Stanford Encyclopedia of Philosophy* (Winter 2010 Edition), edited by E. N. Zalta.
URL = http://plato.stanford.edu/archives/win2010/entries/time/
Mellor, D. H. (1998), *Real Time II*, London: Routledge.
Merricks, T. (2007), *Truth and Ontology*, Oxford: Oxford University Press.
Oaklander, L. N. (2003), "Two Versions of the New Theory of B-Language", in *Time, Tense and Reference*, edited by A. Jokic and Q. Smith, Cambridge, Mass.: MIT Pressh, 271–303.
Oaklander, L. N. (2015), "Temporal Phenomena, Ontology and the R-theory", in *Metaphysica*: 16, 253–269.
Orilia, F. (2012), *Filosofia del Tempo. Il Dibattito contemporaneo*, Roma: Carocci.
Orilia, F. (2014), "Two metaphysical perspectives on the duration of the present", in *Debates in the Metaphysics of Time*, edited by L. N. Oaklander, London: Bloomsbury, 51–70.
Orilia, F. (2016), "Moderate Presentism", in *Philosophical Studies*: 173, 589–607.
Orilia, F. and Oaklander, L. N. (2015), "Do We Really Need a New B-Theory of Time?", in *Topoi*: 34, 157–170.
Prior, A. N. (1959), "Thank Goodness That's Over", in *Philosophy*: 34, 12–17.

15 I wish to thank for their valuable comments Ernesto Graziani, Tomis Kapitan, Greg Landini, Nathan Oaklander, Michele Paolini Paoletti and the above mentioned anonymous referee.

Prior, A. N. (1996), "Some Free Thinking About Time", in *Logic and Reality. Essays on the Legacy of Arthur Prior*, edited by J. Copeland, Oxford: Clarendon Press, 271–303.
Rawls, J. B. (1971), *A Theory of Justice*, Cambridge, MA: Harvard University Press.
Schlesinger, G. (1980), *Aspects of Time*, Indianapolis: Hackett.
Sider, T., Hawthorne, J. and Zimmerman, D. W. (eds.) (2008), *Contemporary Debates in Metaphysics*, Oxford: Blackwell.
Smart, J. J. C. (2008), "The Tenseless Theory of Time", in Sider et al. (2008), 226–238.
Turri, J. (2013), "That's Outrageous", in *Theoria*: 79, 167–171.
Williamson, T. (2002), "Necessary Existents", in *Logic, Thought and Language*, edited by A. O'Hear, Cambridge: Cambridge University Press, 233–251.
Zimmerman, D. W. (2008), 'The Privileged Present: Defending an "A-theory" of Time", in Sider et al. (2008), 211–225.

Elisa Paganini

McTaggart, Lewis and the Problem of Temporary Intrinsics

McTaggart's Paradox is notoriously hard to understand: different interpretations have been given of it and it has even been argued that it has no coherent interpretation.[1] My concern is not to establish whether there is or is not a correct interpretation of the paradox, but to consider whether a particular interpretation is actually an adequate account of a paradox of the passage of time, without considering whether the paradox is really McTaggart's paradox or not. The interpretation I have in mind depends on the assumption that McTaggart's Paradox is a special case of Lewis's Problem of Temporary Intrinsics.[2]

I have two main targets. First, I will point out that the Problem of Temporary Intrinsics cannot simply be applied to the problem of the passage of time and the reason for this is of use for highlighting the difference between the change over time (i.e. the subject of Lewis's Problem) and the change (or passage) of time (i.e. the subject of McTaggart's Paradox).

Once the difference between the two problems have been pointed out, my second aim will be to show that there is a way to present a Paradox of the passage of time in which some differences remain with respect to the Problem of Temporary Intrinsics.

My work is organized as follows: first, I give a presentation of the Problem of Temporary Intrinsics, second, I show that this argument cannot simply be applied to the problem of the passage of time, third, I will present a paradox for the passage of time which is not simply a special case of the Problem of Temporary Intrinsics.

1 See for example Broad (1938) and Dummett (1960) for different interpretations of McTaggart's Paradox (to be found in McTaggart (1908)). See Thomson (2001) for arguments against a coherent interpretation of McTaggart's Paradox.

2 The Problem of Temporary Intrinsics first appeared in Lewis (1986). The hypothesis that McTaggart's Paradox is a special case of Lewis's Problem of Temporary Intrinsics has been first proposed by Craig (1998), and then approved by Rea (2003) and more recently by Rettler (2012).

https://doi.org/9783110594164-011

1 Lewis's Problem of Temporary Intrinsics

Lewis's Problem of Temporary Intrinsics has been very widely discussed since its first publication in 1986.[3] I believe that the problem depends on two pre-theoretical hypotheses we make about things changing over time: the first is that one and the same changing thing exists at different instants of time; the second is that any property characterizing a thing changing is both temporary (i.e. it lasts for a period of time shorter than the entire existence of the changing thing) and intrinsic (i.e. it is possessed by the changing thing independently of any relation it may have with anything else).

In my opinion, the two pre-theoretic hypotheses can be schematically expressed as follows:

1) One and the same changing object (or event) *O* exists at different times.
2) Any property *P* characterizing *O*'s change is exemplified both temporarily and intrinsically.[4]

It may be interesting to note that when we make the second hypothesis pre-theoretically, the first is already incorporated in it, i.e., it is assumed that *O* in 2) is an object (or event) which exists at different instants of time.

The Problem of Temporary Intrinsics depends on the fact that the two hypotheses give rise to a contradiction. The argument showing the contradiction may be schematically presented as follows:

1 One and the same *O* exists at *t* and at *t**. [assumption]
2 Being bent is a property which characterizes a change in *O*, *O* is bent at *t* and *O* is straight (or not bent) at *t**. [assumption]
3 Therefore, one and the same *O* is intrinsically bent and is intrinsically straight (or not bent).

---

3 It is well beyond the purpose of this work to account for the literature on Lewis's Problem of Temporary Intrinsics. It may be useful just to remind that Lewis answered to objections to his argument in two publications: Lewis (1988) and Lewis (2002).

4 Lewis wrote about temporary intrinsic properties. Some objectors to Lewis's Problem of Temporary Intrinsics pointed out that the Problem may be solved if we assume that the properties characterizing an object's or an event's change are instantiated either relationally (see Johnston (1987) and Rettler (2012)) or intrinsically (see Lowe (1988) and Haslanger (1989)). This debate presupposes a subtle distinction between the temporary or intrinsic nature of *properties* and the temporary or intrinsic nature of *property instantiation*. I try to be neutral with respect to this subtle distinction.

Once 1 and 2 are assumed and the two pre-theoretical hypotheses 1) and 2) are accepted, conclusion 3 follows. Conclusion 3 is a clear contradiction and requires the revision of at least one hypothesis grounding it. Lewis himself presents three solutions to the problem envisaged above. Each of the three solutions requires us to revise our image of what it means for something to change over time and the philosophical literature has long discussed which solution is best. I am going to present very briefly the three solutions to the Problem of Temporary Intrinsics: my aim is just to provide the instruments for understanding why I think that the paradox of the passage of time cannot simply be a special case of the Problem of Temporary Intrinsics.

1.1 The first solution to the Problem

The first solution – the one actually defended by Lewis – is to assume that nothing exists in its entirety at different instants of time (i.e. the solution is to deny the first pre-theoretical hypothesis); according to this approach to the problem, what we commonly consider to be objects and events extended in time are constituted by temporal parts, each part being instantaneous and different from any other part. To use the terminology introduced by Lewis, things and events "perdure", being constituted by temporal parts and not being wholly present at each instant of time. If we adopt the theory of temporal parts, the Problem of Temporary Intrinsics disappears as long as there is not something maintaining its identity through time and undergoing a change over time.

The first solution does not deny the second hypothesis, which is to be considered vacuously true. In order to see this, it may be useful to consider that we consider the following assertion vacuously true: any fountain of youth rejuvenates whoever drinks water from it. As long as it is commonly believed there are no fountains of youth, it is taken for granted that the sentence is vacuously true. In the same way, as long as – according to the first solution – it is not the case that one and the same object (or event) exists at different instants of time, there are no properties characterizing a changing thing existing in its entirety at different instants of time, and the second hypothesis is therefore to be considered vacuously true.

1.2 The second solution to the Problem

The second solution is to hold hypothesis 1), i.e. that something exists in its entirety at different instants of time, but to reject hypothesis 2). It is claimed that something existing in its entirety at different instants of time may change over time by having different temporary properties, without assuming that the properties characterizing something's change should be intrinsic.

For example, according to the second solution, I am something existing in its entirety at different instants of time. Suppose, moreover, that I am seated at *t* and that I am not seated at *t**, then – according to the second solution – I have the relational property being-seated-at-*t* and I do not have the relational property being-seated-at-*t**. The two properties are different relational properties and therefore there is no contradiction in assuming that I have one and not the other.

1.3 The third solution to the Problem

The third solution is again a way to reject the first hypothesis, and to allow the second hypothesis to be trivially true. Even if the third solution's approach towards the two hypotheses is equivalent to the first solution's approach, the reasons grounding it are not at all similar.

While the supporter of the first solution maintains that what we commonly consider an object or event is constituted by temporal parts, the supporter of the third solution – the presentist, according to Lewis – does not maintain that there are temporal parts, she claims instead that there is only one genuine time – i.e., the present – and therefore anything existing exists at it.

As in the case of the first solution, the second hypothesis is to be considered trivially true. The third solution excludes things existing at different instants of time and changing in it, these claims are enough to consider the second hypothesis trivially true. As long as nothing maintain its existence at different instants of time, the second hypothesis is considered trivially true.

2 Lewis's Problem and the Change (or Passage) of Time

Let us now try to apply the Problem of Temporary Intrinsics to the passage of time. Just as we believe pre-theoretically that objects and events persisting in time have temporary intrinsic properties (the assumptions which gave rise to Lewis's Problem), we may illegitimately presume the pre-theoretical belief that at least events persisting in time have temporary intrinsic *temporal* properties (by temporal properties I mean the properties "being present", "being future" and "being past"). And we may also presume that the latter pre-theoretic assumption gives rise to a

Paradox which parallels the Problem of Temporary Intrinsics. In this section of my work, I want to argue that this parallelism is not adequate and that this fact may be useful for understanding an important difference between change over time and change of time.

Let us first try to apply Lewis's Problem of Temporary Intrinsics to the problem of temporal change. We seem to adopt the following two hypotheses:

1) One and the same event *E* undergoing temporal change exists at different times.
2) Any temporal property *T* characterizing *E*'s temporal change is exemplified both temporarily and intrinsically.

A moment's reflection shows that we are not at all pre-theoretically disposed to accept the first hypothesis. Let us consider why. Let us suppose that an event *E* is instantaneous, i.e., it exists at a single instant of time. We still suppose that such an event undergoes a temporal change: it passes from being future, to being present and then past. The first hypothesis is therefore not pre-theoretically required in order to account for the passage of an event from being past to being present and from being present to being future.

It may be useful to reflect on the reason why the Problem of Temporary Intrinsics cannot simply be applied to the passage of time and in particular why the first hypothesis is not adequate. In my opinion, while we pre-theoretically assume that any property characterizing change over time pertains primarily to objects or events existing at different instants of time, we pre-theoretically accept that the properties "being past", "being present" and "being future" concern primarily instants of time, which are by definition instantaneous.

3 The Change of Time

If my observation is correct, we pre-theoretically assume that while "being past", "being present" and "being future" pertain to instants of time and only indirectly to events or objects (instantiated at these instants of time), the other properties pertain to events or objects.

The problem of the passage of time may therefore be described as a problem concerning instants of time (and only indirectly objects or events): the problem may be described as the inability of an instant of time to instantiate the properties "being past", "being present" and "being future" both intrinsically and temporarily.

The paradox of the passage of time is not therefore a simple reproduction of the problem of temporary intrinsic change since the temporal properties pertain primarily to instants of time and not to objects and events; moreover instants of time changing their temporal properties are by definition without temporal duration, i.e., they are instantaneous, while objects and events changing over time have temporal duration.

Now, these observations concerning the objects which instantiate temporal properties make the first hypothesis considered (i.e., 1) in §2) inadequate, not only because the object of temporal properties are supposed to be instants of time, but also because they are not assumed to persist in time. We cannot simply assume that "one and the same instant of time exists at different times", an instant of time does not exist at any instant of time different from itself. In order for an instant of time to change its temporal properties, it seems that it should be assumed that it maintains its identity through time, the first hypothesis is therefore to be changed as follows:

1*) Every instant of time *t* maintains its identity through time.

The second hypothesis is that an instant of time instantiates temporal properties both temporarily (i.e., it has them at certain times and not at others) and intrinsically (i.e., it has them independently of any relation it has with any other instant of time). The second pre-theoretic hypothesis may therefore be expressed as follows:

2*) Any temporal property *T* characterizing *t*'s change is exemplified both temporarily and intrinsically.

1*) and 2*) give rise to a contradiction. The argument may be schematically represented as follows:

1 *t* is identical with itself at any instant of time. [assumption]
2 Being present is a temporal property which characterizes *t*'s change, *t* is present at *t* and *t* is not present at *t*'. [assumption]
3 Therefore, one and the same *t* is intrinsically present and is intrinsically not present.

Once 1 and 2 are assumed and the two pre-theoretic hypotheses 1*) and 2*) are accepted, the contradictory conclusion 3 follows. Once again, some of our pre-theoretic assumptions are to be revised in order to avoid the contradiction.

I see three options which may be adopted by whoever wants to avoid the contradiction. The first option is obviously to reject hypothesis 1*). But what is the reason for rejecting it? It is interesting to note that a solution similar to the first solution to the Problem of Temporary Intrinsics cannot be reproduced here. It does not make sense to say that an instant of time has temporal parts. We cannot therefore say that an instant of time does not maintain its identity through time because it has different temporal parts.

A presentist may reject the first pre-theoretic hypothesis by claiming that an instant of time exists only at the present time without maintaining its identity through time. Even if this is a viable alternative, it has not been adopted as far as I know. Different philosophers maintain different theories concerning the identity of instants of time: some philosophers assume that instants of time exist eternally or atemporally and therefore they maintain their identity through time,[5] others assume that they have counterparts as ersatz worlds[6] or they exist in the mind of God[7] and this is what allows them to maintain their identity through time.

If assumption 1*) is accepted, the only way to avoid the paradox is to deny 2*). 2*) can be denied in principle by adopting two different strategies. It can be maintained (and this is the first strategy) that any temporal property characterizing an instant of time is temporary without being intrinsic. Or it can be maintained (and this is the second strategy) that any temporal property characterizing an instant of time is intrinsic without being temporary. The two strategies obviously avoid the paradox; it is an important and difficult philosophical problem to establish whether either of the two solutions accounts for the passage of time or whether they solve the paradox at the cost of denying the passage of time. I am not going to consider this problem here, my concern is the difference between the change in objects or events over time and the passage of time.[8]

For the present occasion, let me observe that the proposed problem of the passage of time is not a simple application of Lewis's Problem of Temporary Intrinsics. First, the two problems pre-theoretically concern different subjects, in one case they are pre-theoretically believed to concern objects and events existing at different instant of time, in the other case they are pre-theoretically believed to concern instants of time, which are instantaneous by definition. Moreover the solutions to the two problems are quite different: Lewis's solution to the problem of temporary intrinsics cannot be applied to the problem of the passage of time, moreover the presentist solution to the problem of temporary intrinsics is not actually applied in the case of the problem of temporal change. And the difference between the solutions to the two problems is a clear indication of the difference between the two problems themselves.

---

5 This is usually maintaied by supporters of the B-Theory of time (see for example Mellor (1981)) or of the hybrid A-B Theory (see for example Smith (2003)).

6 This is suggested by Lewis (1986), and it is endorsed for example by Crisp (2007).

7 See for example Rhoda (2009).

8 I discussed the two solutions to the paradox of time so conceived in Paganini (2005).

Bibliography

Broad, C. D. (1938), *Examination of McTaggart's Philosophy*, Vol. II, Part I, Cambridge: Cambridge University Press.
Craig, W. L. (1998), "McTaggart's paradox and the Problem of Temporary Intrinsics", in *Analysis*: 58(2), 122–127.
Crisp, T. (2007), "Presentism and the grounding objection", in *Noûs*: 41, 90–109.
Dummett, M. (1960), "A Defense of McTaggart's proof of the unreality of time", in *Philosophical Review*: 69(4), 497–504.
Haslanger, S. (1989), "Endurance and temporary intrinsics", in *Analysis*: 49, 119–125.
Johnston, M. (1987), "Is there a problem about persistence?", in *Aristotelian Society Supplementary*: 61, 107–135.
Lewis, D. (1986), *On the Plurality of Worlds*, Oxford, UK: Basil Blackwell.
Lewis, D. (1988), "Rearrangement of particles: Reply to Lowe", in *Analysis*: 48(2), 65–72.
Lewis, D. (2002), "Tensing the Copula", in *Mind*: 111(441), 1–14.
Lowe, E. J. (1988), "The Problems of Intrinsic Change: Rejoinder to Lewis", in *Analysis*: 48(2), 72–77.
McTaggart, J. M. E. (1908), "The unreality of time", in *Mind*: 17, 456–473.
Mellor, D. H. (1981), *Real Time*, Cambridge: Cambridge UP.
Paganini, E. (2005), "Temporal becoming: can we do without it?", in *Philosophical Writings*: 28, 35–44.
Rea, M. C. (2003), "Four-dimensionalism", in *Oxford Handbook of Metaphysics*, edited by M. J. Loux and D. W. Zimmerman, Oxford, UK: Oxford University Press, 246–280.
Rettler, B. (2012), "McTaggart and indexing the copula", in *Philosophical Studies*: 158, 431–434.
Rhoda, A. R. (2009), "Presentism, truth-makers and God", in *Pacific Philosophical Quarterly*: 90, 41–62.
Smith, Q. (2003), "Reference to the past and future", in *Time, Tense and Reference*, edited by A. Jokic and Q. Smith, Cambridge, Mass.: MIT Press.
Thomson, J. J. (2001), "McTaggart on time", in *Philosophical Perspectives*: 15, 229–252.

Alfredo Tomasetta

Is Dualism Compatible with Classical Theism?

1 Introduction

In this paper I am going to discuss the relation between (some forms of) dualism in the philosophy of mind and Classical Theism, focusing on a rather simple point, which, as far as I know, has not been noticed.

The key point of the paper can be briefly stated as follows: a number of important and influential contemporary dualists in the philosophy of mind – whose names are associated with notable arguments against physicalism – subscribe, more or less explicitly, to a thesis that is incompatible with classical theistic belief.

This being the main point of the talk, I'm going to proceed in the following way:

(1) First, I state the thesis I have alluded to – which I label 'Thesis T'.

(2) Second, I show that Thesis T is explicitly endorsed (or, at the very least, clearly suggested) by important and influential contemporary dualists in the philosophy of mind.

(3) Third, I present a short argument to the effect that Thesis T and Classical Theism are indeed incompatible.

(4) Fourth, I conclude by exploring the logical space of possible reactions to the argument presented at point three.

Before starting with point one, a terminological note. I have used, and I will use, the word 'dualism' and cognate expressions in a broad and somewhat idiosyncratic sense, to include not just dualism in the strict sense of the term but also all other anti-physicalist positions in the philosophy of mind – such as, for example, Russellian Monism or Idealism. So why not just use 'anti-physicalism' or 'non-physicalism'? Well, these words simply sound too long and too clumsy to my ears; and perhaps to the reader's as well.

So let us now start with the first point, which is concerned with the statement of what I call 'Thesis T'.

https://doi.org/9783110594164-012

2 'Thesis T'

Thesis T is really the conjunction of two related theses, TN and TS. Thesis TN says that experiencing a conscious state E is a necessary condition for knowing E's nature or essence, while TS says that experiencing a conscious state E – along with appropriate conceptual resources – is sufficient to know E's nature or essence. Here are the theses:

Thesis T

(TN) If one knows what a conscious experience E is (if one knows its nature or essence), then one has (one experiences/is acquainted with) E.

(TS) If one has (experiences/is acquainted with) a conscious experience E – and one has appropriate conceptual resources –, then one knows (or is at least able to know) what experience E is (E's nature or essence).

(As an aside, notice that those who are inclined to accept the first thesis are also usually inclined to say – on reflection – that TN is only a rough and ready statement of a more nuanced principle. For example, they would usually say, on reflection, that one might be able to imagine, and know, the nature of a particular colour experience one has never had, *if one has had some appropriately similar colour experience*. So even those who lean towards thesis TN usually think that there are (or there may be) cases in which one knows what an experience E is without having had it. Having said that, I am nonetheless going to stick to TN: nothing really important will hinge on this point).

TN and TS are rather strong theses, and indeed, to many, very dubious ones. Here, however, I am not concerned with their defence. Rather, I just want to show that a number of influential contemporary dualists subscribe to both TN and TS – but, notice, what will really count (in paragraph four), is their endorsement of TN. Let us therefore move to the second point of the paper

3 TN and TS are endorsed by influential contemporary dualists

I have just said that influential contemporary dualists in the philosophy of mind – whose names are quite naturally associated with well-known arguments against physicalism – endorse TN and TS. Let us now review some of these names – and the associated arguments – and let us see how these philosophers are committed to Thesis T.

3.1 Chalmers (The conceivability argument)

I start with the conceivability argument against physicalism, and focus on the version of it offered by its best-known champion: David Chalmers. The conceivability argument, in its naïve form, just says that an experience-less physical duplicate of our world is conceivable – and that what is conceivable is metaphysically possible; therefore, the argument concludes, experience is not metaphysically determined by physical reality, and physicalism is false. Now, in Chalmers's hands this simple-minded, and quite certainly flawed, argument has become a more effective and very complex piece of philosophical machinery, involving – among its many components – the admission of two kinds of intensions and a related neo-Fregean philosophy of language.[1] I cannot even try to present here the details of this rather baroque and powerful argument. So I shall rest content with highlighting the elements of it – and around it – that are relevant in this context.[2]

(TS) Chalmers

Let us first see how some of Chalmers's ideas lead to the TS thesis. And let us see this in three stages.

First stage – According to Chalmers (2003), we are acquainted with our conscious/experiential states; let us focus on the experience of pain, as many philosophers of mind like to do. If one is acquainted with pain, and one has appropriate conceptual resources/capabilities, then – Chalmers says – one is able to form the phenomenal concept "pain".

Second stage – If one has the concept "water", Chalmers (2006, 2010) holds, one knows a priori, in virtue of one's conceptual competence, what the 'superficial'/apparent properties of water are: according to Chalmers, if you master the concept "water", then you know – *a priori*, and among other things – that water is a transparent, drinkable liquid. Exactly the same holds for phenomenal/experiential concepts such as, for example, "pain". If you master the concept "pain", then you know *a priori* what the 'superficial'/apparent properties of pain are.

1 See, Chalmers (2010).

2 The occasional simplifications will do no harm, because the gist of Chalmers's ideas, I think, is faithfully preserved.

Third stage – The 'superficial'/apparent properties of pain exhaust the reality of pain: as Kripke (1980) has noted, there is no distinction between appearance and reality in the case of conscious experiences.

So we have the following: if we are acquainted with pain and have appropriate conceptual resources, we are able to form the phenomenal concept "pain". In this way we have *a priori* knowledge of pain's apparent properties – that is, of all pain's properties. So we are able to know the nature of pain. Therefore, (TS), if we are acquainted with pain, and have appropriate conceptual resources, we (are able to) know the nature of pain.

(TN) Chalmers

Let us now see how some of Chalmers's (Chalmers (2003)) ideas lead to the TN thesis. Suppose you know the nature of a particular experience – let us stick to pain. This being a piece of knowledge that can contribute to propositional knowledge, the knowledge of the nature of pain you have requires that you have the phenomenal concept "pain" (it is a *phenomenal* concept given that pain is a *phenomenological or experiential* state).

Now: Chalmers gives a 'constitutional' account of phenomenal concepts according to which, roughly, when one possesses a token of a phenomenal concept "E", then an experienced token of a conscious state E (at least partly) constitutes the concept itself. So, if one possesses the phenomenal concept "pain", then an experienced token of pain (at least partly) constitutes the phenomenal concept "pain" itself. Therefore, if one possesses the phenomenal concept "pain", then one experiences pain.

Hence we have that if you know the nature of pain, then you have the phenomenal concept "pain", and that if you have the phenomenal concept "pain" then you experience pain. Therefore, (TN), if you know the nature of an experience (of, say, pain), you have that very experience.

3.2 Jackson and Russell (The knowledge argument)

We have just seen that David Chalmers is committed to Thesis T. The same holds for the philosophers associated with the so-called "Knowledge Argument", which is really not a single argument but rather a family of arguments that has a long story, tracing back to C. D. Broad and even further back to Samuel Alexander. One can find more recent versions of the argument, for example, in Russell (1927), Feigl (1958), Nagel (1974), and of course Jackson (1982, 1986). For the sake of brevity, I focus only on Jackson's and Russell's versions of the argument.

As for Jackson, everybody knows very well the case of Mary, the super-duper neuroscientist confined in a black and white room who specialises in colour vision and knows all of the physical facts about it. She knows, for example, all there is to know about the surface reflectance properties of physical objects, wavelengths of light, and retinal stimulation, but she has never actually experienced any colours other than black, white and shades of grey. Jackson then asks: "What happens if Mary is released from her black and white room? Does she *learn* anything or not?" According to him, when Mary leaves the room and, for the first time, gazes at an object that is red – and that she knows to be red –, she learns *what it's like to see red.* Jackson concludes that, because physicalism requires that all facts are physical facts, physicalism is false.

So much for Jackson's famous version of the knowledge argument. Here is Russell's elegant and very concise version of it (Russell (1927), p. 389): "It is obvious that a man who can see knows things which a blind man cannot know. But a blind man can know the whole of physics. Thus the knowledge which other men have, and he has not, is not part of physics". And now let us see how these arguments are related to Thesis T.

(TS) Jackson

Mary leaves the room and gazes upon a red object that she knows to be red. That is to say: she has the concept "red" and she experiences red. In this condition, Jackson says, Mary learns *what it's like to see red.* Not just, notice, what it is like to be red, but what it is like to have a visual experience of red. That is, she knows what an experience of red is.

Therefore, one may easily conclude, having an experience (of red, for example), and having appropriate conceptual resources, suffices to know what that experience is. And this is indeed the TS Thesis.

(TN) Jackson

Now for the crucial TN – as I have said, TN will play the crucial role in paragraph four. In this case things are a bit less straightforward, but I think it is quite obvious that the Mary case points very strongly towards TN (Indeed, it is not at all unusual to find philosophers discussing the Mary argument and simply asserting that the argument itself implies the truth of TN[3]).

Suppose, first, that Mary knows not just all the physical facts about colour vision, but all physical facts *simpliciter*. With this supposition, of course, the Mary argument would be exactly the same.[4]

Suppose now, drawing on a scenario devised by Paul Churchland (1989a,b), that persons are immaterial souls constituted by some 'ectoplasmic' stuff causally interacting with physical bodies. Suppose, further, that Mary knows all the physical *and the ectoplasmic* facts. Even in this case, Churchland suggests, Mary would learn *something new* upon leaving the room.

I think Churchland is quite right. And I also think that his consideration allows one to grasp the real underlying point of the Mary story. Which is, I maintain, the following.

Mary could know all the physical *and ectoplasmic* facts, and still lack knowledge of what it is like to see red. But she could also know all there is to know about any other sort of fact without knowing what it is like to see red, *provided that her knowledge concerned things she need not instantiate in order to know them*. The real point of the Mary story seems to me clearly to be that even though Mary knew *all that is knowable without being instantiated by the subject who knows it*, she would still not know what it is like to see red. This, I submit, is the guiding intuition underlying the Mary thought-experiment, an intuition that can also be expressed in the following way: 'third personal knowledge' (that is: knowledge concerning things one need not instantiate in order to know them), can never give knowledge of what an experience E is; *the only* way to know the nature of an experience (of red, for example) is by having it; and this idea, of course, is TN thesis.

3.2.1 (TN) Russell

Let us now consider the case of Russell's version of the knowledge argument, focusing, for brevity's sake, only on the crucial TN thesis (but it would not be difficult to argue that Russell endorsed TS as well).

---

3 To give just one example among many, Fred Dretske writes: "There is an argument due to Frank Jackson [i.e., the Mary argument] [...] that if you do not experience colour for yourself then [...] you do not know what it is like to experience red, blue, and other colours. The argument easily generalises [to any other experience]" (Dretske (1995), p. 81). So according to Dretske, Jackson is saying that if you don't experience E, then you don't know what E is. And contraposing, one obtains TN.

4 Jackson himself, in Jackson (1986), presents the Mary argument in this way.

Showing that Russell in fact subscribes to TN is quite a straightforward task. If one is a blind man, Russell says, it is obvious that one lacks some kind of knowledge. What kind of knowledge does one lack? Well, among other things, a man born blind certainly does not know what, for example, an experience of blue is – he does not know what the nature of an experience of blue is. So, if one has no visual experience at all, then one does not know what an experience of blue is. Hence, by contraposition, if one does know what an experience of blue is, one has visual experience. And what kind of visual experience? Well, an experience of blue – what else?[5]

So if one knows what a visual experience of blue is, one has an experience of blue. Generalising: if one knows what a visual experience is, one has that visual experience. And, from this, it is just a short step to generalise further and to conclude that if one knows what an experience E is, one has this very experience. Which is TN thesis.

3.3 Fumerton, Goff, Strawson (The argument from revelation)

Until now we have seen that Chalmers, Jackson[6], the Russell of the Knowledge Argument, and other important supporters of versions of this argument, are committed to Theses TN and TS. Let us now extend our list of supporters of Thesis T by considering other dualist philosophers associated with the so-called 'Argument from Revelation Against Physicalism'.

The basic idea of the argument is, roughly, the following. If one has an experience of, say, pain (and one has appropriate conceptual resources), then this experience *reveals* the nature of pain itself, so that one knows (or at least is able to know) its nature. Subjects having experiences of pain, therefore, know (or are able to know) the nature of pain; and yet these subjects do not know (and are not able to know) that pain has a physical nature. So, the argument concludes, the nature of pain (and of every other experience) is not a physical one.

Dualist arguments based on this kind of reasoning are perhaps less well known than the conceivability and the knowledge arguments. Yet they are increasingly popular, and have recently been put forward, among others, by dua-

5 Or, perhaps, some appropriately similar colour experience – but, as I have said earlier, I overlook this complication.

6 At least as a supporter of the Mary argument – on which he changed his mind, as is well-known.

lists such as Philipp Goff (forthcoming), Richard Fumerton (2013) and Galen Strawson (2006) - and have been critically discussed, for example, by Daniel Stoljar (2009) and Andreas Elpidorou (forthcoming).

Now, it is completely obvious that those committed to the Argument from Revelation are also committed to TS: TS is, of course, the starting point of the argument. But notice that, in articulating the general philosophical framework surrounding their endorsement of TS, such philosophers as Fumerton, Goff and Strawson take TN for granted as well.

Fumerton, for example, supports a foundationalist epistemological theory according to which *the only* way to know the nature of our experiences is to be acquainted with them.

And Strawson's assertion that, in the case of experiences, "the having is the knowing" (Strawson (2006), p. 251) (a thesis whose general spirit is shared by Goff as well) implies that if you know what an experience E is, then you have E – which is TN thesis.

So notable friends of the Argument from Revelation Against Physicalism adopt not just TS but also TN thesis. And their names enlarge the list of TN's influential supporters.

This concludes the second part of the paper. Let us now move to paragraph four, which is devoted to showing the incompatibility between thesis T(N) and Classical Theism.

4 T(N) is incompatible with classical theism

What do I mean by "Classical Theism"? In order to clarify this, let us first make what seems to be a digression, and let us consider the following list of conscious mental states:

pain	rage
anger	melancholy
envy	horror
sexual desire	relief
depression	fear

For brevity's sake, I'll refer collectively to these conscious mental states as "the Ms". And now back to Classical Theism.

I speak of "Classical Theism" having in mind the set of core beliefs (perhaps a set with fuzzy borders, as it were) that are shared by traditional monotheistic theologies such as Christian, Jewish and Islamic theologies, core beliefs concerning a personal, transcendent, God and some of Her/His attributes. All of this is admittedly rather vague, but I do not need to be more precise, given that all I am interested in here are two beliefs which, I am confident, all readers will recognise as part of the core traditional theistic set of beliefs about God – as part of Classical Theism, that is. The two ideas are the following:

A) God is omniscient (She/He knows all that is knowable).

and

B) God does not instantiate the Ms (God is not envious, has no fear, is not melancholic, is not depressed, is not horrified, and so on and so forth).

At this point, it is very easy to show that T(N) and Classical Theism (CT) are indeed incompatible. Let us see.

---

1) Assume that TN and CT are both true.
   If CT is true, then B) is true. So
2) God does not instantiate the Ms.
   By TN (and *modus tollens*), if God does not instantiate the Ms, then
3) God does not know the natures of the Ms.
   But if God does not know the natures of the Ms, then
4) God is not omniscient.
   And yet if, as we have assumed, CT is true, then A) is true as well; that is
5) God is omniscient.
   Contradiction!

So the initial assumption must be rejected: TN and CT cannot both be true. If TN is true, then Classical Theism is not true: T(N) and Classical Theism are incompatible.

5 Exploring possible reactions

How can a theist react to the previous argument? In this last section I consider, and comment upon, three possible options (the comment to the first one involves TS).

Option 1: T, not-A, B

One option for the theist is to retain thesis T (and so retain TN), to subscribe to B whilst renouncing A: having experience E is (roughly) necessary and sufficient to know E's nature; God is not envious, or depressed and so on; but She/He does *not* know everything that is knowable. Perhaps some theists may rest content with this non-classical doctrine.

Yet, one should carefully consider what exactly God's ignorance amounts to in this case. We (beings endowed with appropriate conceptual resources) experience the Ms, and so, by thesis T (that is, by TS), we know the natures of the Ms. God, instead, does not experience the Ms, (by B), so by thesis T (that is, by TN) She/He does not know the natures of the Ms. Hence the supporter of Option 1 is bound to say not just that God is not omniscient, but also that *we human beings know things that God does not know*, namely the natures of some conscious mental states. And this, moreover, implies that God does not have full knowledge of 'our hearts' as many theists would want to say. These, I guess, are sufficient reasons for a theist to be suspicious of Option 1.

Option 2: T, A, not-B

A second option for the theist would be to retain thesis T (and so TN), to subscribe to A whilst renouncing B: in this case, having experience E is (roughly) necessary and sufficient to know E's nature; God does know everything that is knowable, but She/He is envious, or depressed or melancholic or fearful ... and so on. And I think that attributing even *some* of the Ms to God is not really an alternative for the vast majority of theists. So let us set aside the second option as well.

Option 3: not-T (denying TN), A and B

The last option for the theist that I am going to consider is, not surprisingly, the following: endorsing Classical Theism – and so endorsing both A and B –, and rejecting Thesis T denying TN.

This move, of course, forces the classical theist to also reject the positions of such eminent dualists as Chalmers, (the early) Jackson, Russell, Fumerton and so on. "But so what?", one may say: "A classical theist may just be a different kind of dualist!". And yet, I think, in this regard things are not as easy as it may seem. The classical theist adopting Option 3, and wanting to uphold dualism, has to come up with a good argument, A, for dualism. This argument moreover must not involve TN and must be such that the general philosophical framework that the endorsement of A leads to does not involve TN itself. Is there such an argument? This is a challenge to be met. But, possibly, it is a challenge that cannot be met. As Richard Fumerton[7] has recently suggested (if my reading of him is correct), if one rejects Thesis T, then there is no plausible argument for dualism. So *if there is a plausible argument for dualism, then Thesis T has to be accepted*. This is an idea that, for example, is also quite clearly suggested by Fred Dretske in chapter three of Dretske (1995) – and I suggest it to the reader as a reasonable conjecture, awaiting counterexamples.

If the conjecture were true, however, well-argued dualism would imply TN, and so no classical theist denying TN could be a sensible dualist. She/He would instead be a physicalist in the philosophy of mind – and, notice, there are real cases of this combination of views: Peter Van Inwagen is a renowned classical theist *and* a physicalist in the philosophy of mind.[8]

But notice further that the classical theist (who is, or wants to be, a physicalist regarding the mind) cannot think that mental states are identical to physical states: God, according to CT, has mental states, but She/He is not physical. Neither can the classical theist say that mental states are constituted by – without being identical with – some physical or non-physical stuff (as the case may be): God, according to CT, is not constituted by any divine stuff – or by anything at all. So, perhaps, the physicalist classical theist should be a functionalist. And yet functionalism seems to require a complex causal/functional arrangement of internal parts which does not really fit well with the divine simplicity usually asserted by CT.

Thus, it seems as if Classical Theism cannot easily be combined with any metaphysical doctrine concerning the mind. Is this really so? Maybe not, but showing this, it seems to me, is certainly not an easy task.

Bibliography

Chalmers, D. (2003), "The Content and Epistemology of Phenomenal Belief", in *Consciousness: New Philosophical Perspectives*, edited by Q. Smith and A. Jokic, Oxford, Oxford University Press, 220–272.

Chalmers, D. (2006), "Two-Dimensional Semantics", in *The Oxford Handbook of Philosophy of Language*, edited by E. Lepore and B. C. Smith, Oxford, Oxford University Press, 574–606.

---

7 Fumerton (2013), p. 92, footnote 1.

8 See, for example, Van Inwagen (1994, 2014).

Chalmers, D. (2010), "The Two-Dimensional Argument Against Materialism", in *The Character of Consciousness*, edited by D. Chalmers, Oxford, Oxford University Press, 141–205.
Churchland, P. (1989a), "Reduction, Qualia, and the Direct Introspection of Brain States", in *A Neurocomputational Perspective*, edited by P. Churchland, Cambridge, MA: MIT Press, 47–66.
Churchland, P. (1989b), "Knowing Qualia. A Reply to Jackson", in *A Neurocomputational Perspective*, edited by P. Churchland, Cambridge, MA: MIT Press, 67–76.
Dretske, F. (1995), *Naturalizing the Mind*, Cambridge, MA: MIT Press.
Elpidorou, A. (forthcoming), "A Posteriori Physicalism and Introspection", in *Pacific Philosophical Quarterly*.
Feigl, H. (1958), "The 'Mental' and the 'Physcial'", in *Minnesota Studies in the Philosophy of Science*, edited by H. Feigl, G. Maxwell and M. Scriven, Minneapolis: University of Minnesota Press, 370–497.
Fumerton, R. (2013), *Knowledge, Thought, and the Case for Dualism*, Cambridge: Cambridge University Press.
Goff, P. (forthcoming), *Consciousness and Fundamental Reality*.
Jackson, F. (1982), "Epiphenomenal Qualia", in *Philosophical Quarterly*: 32, 127–136.
Jackson, F. (1986), "What Mary Didn't Know", in *Journal of Philosophy*: 83, 291–295.
Kripke, S. (1980), *Naming and Necessity*, Cambridge MA: Harvard University Press.
Nagel, T. (1974), "What Is It Like to Be a Bat?", in *Philosophical Review*: 83, 435–450.
Russell, B. (1927), *Outline of Philosophy*, London: Routledge.
Stoljar, D. (2009), "The Argument from Revelation", in *Conceptual Analysis and Philosophical Naturalism*, edited by D. Braddon-Mitchell and R. Nola, Cambridge, MA: MIT Press, 113–137.
Strawson, G. (2006), "Panpsychism? Reply to Commentators", in *Consciousness and its Place in Nature*, edited by A. Freeman, Exeter: Imprint Academic.
Van Inwagen, P. (1994), "Quam Dilecta", in *God and the Philosophers: The Reconciliation of Faith and Reason*, edited by T. V Morris, New York, Oxford University Press, 31–60.
Van Inwagen, P. (2014), "Causation and the Mental", in *Existence. Essays in Ontology*, by P. Van Inwagen, Cambridge: Cambridge University Press, 238–258.

Kazimierz Trzęsicki

In What Sense is God Infinite? A Consideration from the Historical Perspective

1 Introduction

Throughout history the Infinite has been named variously: the Absolute, the One, the Unconditioned, the Unlimited, the Indivisible, and the Indefinite. Of all that can be attributed to God, infinitude is not only the most difficult to grasp but man is unable even to comprehend God's infinity. The dogma of God's actual infinity is of the greatest import for Christian theology. The dogma is relatively late. The position of the Catholic Church declared in the Fourth Lateran Council (1215), is again clearly stated in the following pronouncement of the Vatican Council (1868–1870): Chapter 1 *On God the creator of all things* of the SESSION 3: 24 April 1870: *Dogmatic Constitution on the Catholic Faith* it is stated that:

> The holy, catholic, apostolic and Roman church believes and acknowledges that there is one true and living God, creator and lord of heaven and earth, almighty, eternal, immeasurable, incomprehensible, infinite in will, understanding and every perfection. (http://www.papalencyclicals.net/Councils/ecum20.htm)

The term "infinity" applies primarily to time, space, quantity, and magnitude; in its derived meaning it is applied to perfection, e.g., wisdom and power. The concept of infinity is important in mathematics and physical science, in philosophy and theology. Moreover, though in any of these disciplines the concepts differ one from another, a deeper understanding of one needs an understanding of the others. There was an important interaction between the development of philosophy, theology, mathematics, physical science, and the development of the idea of infinity. Concurrent contributions and cooperation between these disciplines resulted in diverse concepts of infinity. Views of what the infinite is were a source of progress as well as errors in each of these disciplines. Philosophers and theologians were always fascinated with infinity. Mathematicians tried to avoid it or even demonstrated hostility. Nevertheless, already ancient disputes over infinity are full of references to mathematics (Knorr (1982)). Moreover, the question of infinity has played and still plays a key role in the conceptual development of mathematics. Though infinity features in the calculus of Newton and Leibniz, it was not rigorously defined until the late 1800s, and even in the 19th century many mathe-

https://doi.org/9783110594164-013

maticians found infinity to be vaguely distasteful. For, e.g. Gauss it was only *façon de parler*.

A connection between infinity and God exists in nearly all religions. The question in what sense is God infinite is as old as the infiniteness that has been attributed to Him. What do we mean, e.g., when we attribute infinite goodness or power to God? It seems that there is no other attribute of God more known to theologians than infinity (Drozdek (1995), p. 127). There are two disparate trends in theology: one that tries to find an answer to the problems that come from God's infinity, and one that insists on the impenetrable mystery of the infinite (Le Blanc (1993), p. 51). We shall endeavor to state what philosophers meant when they said that God is infinite in his being or perfection. Special attention will be given to philosophers who have influenced Cantor's conception of infinity. According to Dauben (1979), ch. 6:

> Having dealt with Aristotle and the scholastics, Cantor undertook an investigation of other works by some of the most impressive thinkers of the seventeenth century, a century that witnessed serious and often profound analysis of the nature of infinity. He suggested that anyone interested in such things would do well to consult Locke, Descartes, Spinoza, and Leibniz, while Hobbes and Berkeley were highly recommended as additional reading.

2 Philosophy of infinity in antiquity

From the time people began to think about the world they lived in, questions about infinity arose.[1] Greek philosophers spoke about infinite attributes of *ἀρχή*, the source, origin, or root of things that exist, and – as is the case with Anaximander of Miletus (c.610–c.546 BC) – made it *ἀρχή* itself. An infinite (or limitless) and infinitely divisible primordial form of matter, *ἄπειρον*, gives rise to all natural phenomena, to all finite things.[2] Infinity is indeterminate and limitless, es-

1 The page http://www.logicmuseum.com/cantor/Phil-Infinity.htm. is a selection of philosophical writings about the infinity with links to other resources on the net. For more about surviving classical Greek sayings and writings about infinity see Sweeney (2012) and Sinnige (1968).

2 To the idea of *apeiron* have linked should be after physicists:

> Heisenberg thus arrived at the idea that the elementary particles are to be seen as different manifestations, different quantum states, of one and the same "primordial substance". The elementary particles, it would follow, are the only possible manifestations of matter. Because of its similarity to the primordial substance hypothesized by Anaximander, Born called this substance *apeiron*. (Simonyi (2012), p. 546)

sentially acquiring the negative value of imperfection. The question as to which entities and properties are infinitely large, infinitely small, infinitely divisible, and infinitely numerous is one of the oldest questions of speculative thinking. David Hilbert (1862–1943) said (Hilbert (1925), p. 163):

> Das Unendliche hat wie keine andere Frage von jeher so tief das *Gemüt* der Menschen bewegt; das Uendliche hat wie kaum eine andere *Idee* auf den Verstand so anregend und fruchtbar gewirkt; das Unendliche ist aber auch wie kein anderer *Begriff* so der Aufklärung bedürftig.
>
> No other question has ever moved so profoundly the spirit of man; no other idea has so fruitfully stimulated his intellect; yet no other concept stands in greater need of clarification than that of the infinite. (Newman (1956) and Maor (1987)).

Methodological reflections about infinity are due to Aristotle (385–322 BC). He distinguished between the potential infinite and the actual infinite, an idea that through the centuries would dominate thinking about infinity. The potential infinite is an extension of the finite, constructible by addition or division from the finite by some rule or process that is never in fact completed. Any result of the construction is an actually finite object but capable of incremental increase to infinity. The infinite has being only as a process that can be repeated over and over again without end, but which, at any moment, has only a finite number of components, and thus this infinity exists only potentially. The actual infinite is an object composed of an infinite collection of parts, processes, or elements. The actually infinite entity is not capable of further incremental increase. Based on the paradoxes of the infinite discovered by Zeno of Elea (c.490–430 BC) and others, Aristotle argued against the actual infinite. He and his followers, in order to use the notion of infinity in a coherent manner, maintained that actual infinities could not exist (Aristotle (1930), Book VI). The infinite could not be in any way actual, even as an idea in the mind of a human (Aristotle (1930), III. 7): *Infinitum actu non datur.*[3] Aristotle was aware that such an opinion might be incompatible with mathematics (Torretti (1978), p. 9). He claimed that mathematicians would not require an actual infinite magnitude:

> by disproving the actual existence of the infinite in the direction of increase ... In point of fact they do not need the infinite and do not use it. They postulate only that the finite straight line may be produced as far as they wish. It is possible to have divided in the same ratio as the largest quantity another magnitude of any size you like. Hence, for the purposes of proof,

3 Cantor, considering reasons for Aristotle's rejection of the real existence of the infinite, argued that he presupposed that there were only finite numbers and as a result his argument involved *petitio principi* (Cantor (1932b), ¶, pp. 173–174).

> it makes no difference to them to have such an infinite instead, while its existence would be in the sphere of real magnitudes. (Aristotle (1930), 207 b 27–34)

One can never actually count to infinity. Aristotle rejected the actual existence of the continuum.[4] Thus Zeno's arrow does reach the target because space is not actually infinitely divided. But if there is no actual infinity, how are we to understand the existence of the infinite continuum in geometry? Aristotle's ideas had widespread influence for many centuries; in particular, his opinion was convincing for Hellenistic mathematicians, e.g. for Euclid (323–283 BC.). In his *Elements* (Book IX, Proposition 20) Euclid wrote that there are more prime numbers than contained in any collection, i.e. that there was a potential infinite collection of these numbers rather than that the collection of prime numbers was itself infinite. The idea of the potential infinity of a continuum inspired a method of exhaustion, that e.g. was extensively used by Archimedes (287–212 BC) to arrive at arithmetic formulas pertaining to geometric figures.[5] But Archimedes, as readings of the Archimedes Palimpsest hints, at least had an intuition about actual infinite quantities (https://en.wikipedia.org/wiki/Archimedes_Palimpsest). Aristotle's conception of the infinite plays an important role in contemporary disputes, and many philosophers still look to him for guidance and inspiration in many areas.

Ancient Greeks, and Plato himself, conceived infinity as an unlimited, unbounded, indefinite, unintelligible chaos. The concept of the infinite had negative connotations, in Aristotle's words: being infinite is a privation, not a perfection, but the absence of a limit.

Plotinus (204–270), a neo-Platonist philosopher, was largely influenced by Philo. Plotinus was one of the first important Greek philosophers to help change the conception of infinity as something imperfect. Agreeing with Aristotle that

4 Because of the impossibility of expressing geometric magnitudes in terms of arithmetic quantities, the Greeks considered arithmetic and geometry to be fundamentally distinct.

5 By generalizing Eudoxus' technique to arbitrary curves and using analytic geometry – the Greeks considered arithmetic and geometry to be fundamentally distinct – to apply the principles to algebraic equations, G. W. Leibniz (1646–1716) and Sir Isaac Newton (1642–1727), created what is now called calculus. For McFarlane:

> Here is perhaps the most powerful example of how metaphysical thought about the Infinite has had dramatic consequences in the world. Their techniques involved inconsistencies because Newton and Leibniz, following in Cusa's footsteps, granted actual existence to the infinitely large and infinitely small, and by admitting that an infinite progression can result in an actual limit. http://integralscience.org/cusa.html

there is no actual infinite in the sensible realm, Plotinus asserted the reality of the actual Infinite in a transcendent realm known only through mystical insight. St. Augustine was acquainted with Plotinus' thought.

By the end of antiquity, Augustine of Hippo (354–430), an early Christian theologian and philosopher whose writings influenced the development of Christianity and western philosophy, merged much of Platonism and Neoplatonism with Christianity, and with the Bible's concept of the infinity of God. According to him God is greater than infinity. Humans, as finite beings, are not able to comprehend infinity, and, it is a famous maxim of Augustine:

> si comprehendis non est Deus.
> that which is comprehended by you is not God.

In Augustine of Hippo (1887) he argued *Against Those Who Assert that Things that are Infinite Cannot Be Comprehended by the Knowledge of God*. He explained:

> And thus, if everything which is comprehended is defined or made finite by the comprehension of him who knows it, then all infinity is in some ineffable way made finite to God, for it is comprehensible by His knowledge. Wherefore, if the infinity of numbers cannot be infinite to the knowledge of God, by which it is comprehended, what are we poor creatures that we should presume to fix limits to His knowledge, and say that unless the same temporal thing be repeated by the same periodic revolutions, God cannot either foreknow His creatures that He may make them, or know them when He has made them? God, whose knowledge is simply manifold, and uniform in its variety, comprehends all incomprehensibles with so incomprehensible a comprehension, that though He willed always to make His later works novel and unlike what went before them, He could not produce them without order and foresight, nor conceive them suddenly, but by His eternal foreknowledge. (Augustine of Hippo (1887), Book XII, ch. 18)

God recognizes even Himself as a finite being. Only God due to His infinite nature is able to recognize infinity on its own and only He is capable of infinite thoughts. This is in accordance with an earlier thought of Boethius (c. 480–524). In Boethius (2007) he distinguished between the eternal and perpetual (everlasting):

> ...Platonem sequentes deum quidem aeternum, mundum uero dicamus esse perpetuum. http://faculty.georgetown.edu/jod/boethius/jkok/5p6_t.htm
>
> ...following Plato let us say that God is in fact eternal, while the universe is perpetual. (Boethius (2007), Book V and VI)

For God eternity is present:

> Quoniam igitur omne iudicium secundum sui naturam quae sibi subiecta sunt comprehendit, est autem deo semper aeternus ac praesentarius status, scientia quoque eius omnem temporis supergressa motionem in suae manet simplicitate praesentiae infinitaque praeter-

iti ac futuri spatia complectens omnia quasi iam gerantur in sua simplici cognitione considerat. (http://faculty.georgetown.edu/jod/boethius/jkok/5p6_t.htm)

Since, then, every mode of judgment comprehends its objects conformably to its own nature, and since God abides for ever in an eternal present, His knowledge, also transcending all movement of time, dwells in the simplicity of its own changeless present, and, embracing the whole infinite sweep of the past and of the future, contemplates all that falls within its simple cognition as if it were now taking place. (Boethius (2007), Book V and VI)

According to Augustine, without knowledge of mathematics “our mind could not bear the great light” of the knowledge of God (Drozdek (1995), p. 128). Numbers are also important to comprehending reality (*ibid*, p. 129). “Knowledge of numbers is a path leading to God, ...” (*ibid*, p. 130). The Augustine conception of infinity in theology had the strongest impact on Georg Cantor, the founder of set theory (*ibid*, p. 136).

3 Infinity of God in the Middle Ages

In the medieval period the question of the intelligibility and conceivability of infinity was disputed mainly for theological reasons. Infinity had come to be understood as something endless, unlimited, and immeasurable, but not necessarily chaotic, as was the case in the ancient era. Various mathematical arguments were being devised that helped develop a more subtle understanding of the infinite as represented in geometrical continuity. Scholastic writings on infinity were studied by Cantor.

To Anselm of Canterbury (1033–1109) the reason the human soul does not see God directly is twofold, stemming both from finite human nature and from infinite divine nature.

Utiq[ue] & obscuratur sua brevitate, & obruitur tua immensitate.
Without doubt it is both obscured by its smallness and overshadowed by Your immensity. (Logan (2009), p. 46)

To Anselm God is the source of every true thing that gives light to the rational mind, in which everything which is true exists, and outside which there is only nothing and falsehood, which sees in one glance whatever things have been made, and by whom and through whom and how they were made from nothing (Logan (2009), 14.8, p. 46). There must be a nature that does not have a superior, otherwise the gradations would be infinite and unbounded, which he considers absurd.

Ergo d[omi]ne non solum es quo maius cogitari nequit; sed es quiddam maius qua[m] cogitari possit. Q[uonia]m nanq[ue] valet cogitari e[ss]e aliquid huiusmodi; si tu non es hoc ipsum potest cogitari aliquid maius te, q[uo]d fieri nequit.

Therefore, Lord, not only are You [something] than which a greater cannot be thought, but You are also something greater than can be thought. For, since it can be thought that there is something of this kind, if You are not this very thing, it is possible to think of something greater than You – and this cannot be done. (Logan (2009), 15.2, p. 46)

God is not only *maior quam cogitari potest*, greater than can be thought, but such that human understanding cannot attain it. He is a supreme and inaccessible light which shines too brightly (Logan (2009), 16.4). God is wholly and always. He does not exist in space or time. There are no parts of God. He is without any spatial extension. God is always present. God is entirely present whenever and wherever God is.

…; tu solus incircumscripti & [a]eternus.
…, You alone are unlimited and eternal. (Logan (2009), 13.2, p. 46)

Anselm's characteristic of God is purely negative, i.e. saying He is immutable is saying that He does not change. It is typical: attributing infinity to God and negative theology[6] go hand in hand.

Platonic insight continued to dominate the Christian worldview until St. Thomas Aquinas (1224–1274), widely considered the Catholic Church's greatest theologian, the *angelic doctor*. Thomas did have respect for Aristotle the Philosopher. He grounded his philosophical and theological thought in Aristotelian philosophy, in particular in his concept of infinity, which plays a crucial role in his philosophy and theology. Aristotle deals with infinity in the context of prime unmoved mover. He argued that this prime unmoved mover was of necessity without magnitude or parts, and was infinite (Aristotle (1930), VIII, 10).

The Christian God is infinite. From the philosophical point of view, Aquinas' Christian God is a counterpart of the Aristotelian prime unmoved mover. According to Aristotle everything infinite is imperfect. Finite and infinite belong to quantity or to what is finite according to place. Neither an actually infinite multitude[7] nor an actually infinite magnitude is possible. But there is no quantity in God and He is not a body. According to Thomas:

6 Apophatic theology is characteristic of orthodox churches. See, e.g. Staniloae (1994), p. 134.

7 In the essay *De Aeternitate Mundi* (1270) he said that no proof that an infinite multitude had yet been given. In the *Questiones Quodlibetales* (IX q1, XII a2) he suggests that an actually infinite multitude may be possible.

> ...omnes antiqui philosophi attribuunt infinitum primo principio, ut dicitur in III Physic., et hoc rationabiliter, considerantes res effluere a primo principio in infinitum. Sed quia quidam erraverunt circa naturam primi principii, consequens fuit ut errarent circa infinitatem ipsius. Quia enim ponebant primum principium materiam, consequenter attribuerunt primo principio infinitatem materialem; dicentes aliquod corpus infinitum esse primum principium rerum.
>
> All the ancient philosophers attribute infinitude to the first principle, as is said (Phys. iii), and with reason; for they considered that things flow forth infinitely from the first principle. But because some erred concerning the nature of the first principle, as a consequence they erred also concerning its infinity; forasmuch as they asserted that matter was the first principle; consequently they attributed to the first principle a material infinity to the effect that some infinite body was the first principle of things. (Thomas Aquinas (1947), I a, q.7, a.1)

The question of the infinitude of God is solved by Thomas by introducing the concept of infinity in essence:

> Forma autem non perficitur per materiam, sed magis per eam eius amplitudo contrahitur, unde infinitum secundum quod se tenet ex parte formae non determinatae per materiam, habet rationem perfecti. Illud autem quod est maxime formale omnium, est ipsum esse, ut ex superioribus patet. Cum igitur esse divinum non sit esse receptum in aliquo, sed ipse sit suum esse subsistens, ut supra ostensum est; manifestum est quod ipse Deus sit infinitus et perfectus.
>
> On the other hand, form is not made perfect by matter, but rather is contracted by matter; and hence the infinite, regarded on the part of the form not determined by matter, has the nature of something perfect. Now being is the most formal of all things, as appears from what is shown above (Question [4], Article [1], Objection [3]). Since therefore the divine being is not a being received in anything, but He is His own subsistent being as was shown above (Question [3], Article [4]), it is clear that God Himself is infinite and perfect. (Thomas Aquinas (1920), First Part, Question 7, Article 1)

For Aristotle everything was composed of form and matter. Thomas in his *De Ente et Essentia* (On Being and Essence) added another composition: everything was composed of being and essence. God was simple and the only simple being. The act of existing was other than essence or quiddity. God was a being whose essence was the act of existing. God was (simply) existence only. Essential infinity was nothing else than an infinity of (pure) existence. The questions how the "essential infinity" should be conceived and how it relates to other infinities is the subject matter of considerations by Tapp (2015).[8]

8 https://sites.google.com/site/ontologicalworkshops/ontological-workshop-2014/abstracts, http://www.math.uni-hamburg.de/home/loewe/HiPhI/abstracts.html

The Fourth Lateran Council declared that God is:

> unspeakably elevated above all things that exist, or can be conceived, except Himself.

Thomas says that because God is infinite, we are able only speak of Him by analogy.

> Deus enim omnibus providet secundum quod competit eorum naturae. Est autem naturale homini ut per sensibilia ad intelligibilia veniat, quia omnis nostra cognitio a sensu initium habet. Unde convenienter in sacra Scriptura traduntur nobis spiritualia sub metaphoris corporalium.
>
> Now it is natural to man to attain to intellectual truths through sensible objects, because all our knowledge originates from sense. Hence in Holy Writ, spiritual truths are fittingly taught under the likeness of material things. (Thomas Aquinas (1947), I a, q.1, a.9)

It is proper to say that the creature is like God rather than that God is like the creature. He says:

> Non igitur Deus creaturae assimilatur, sed magis e converso.
> God, then, is not likened to a creature; rather, the converse is true. (5, I, XXIX, 6)

Thomas was an advocate of negative theology.

4 The question of the infinity of God in the modern period

To Nicholas of Cusa (1401–1464), the most important German thinker of the fifteenth century (Jaspers (1964) and Kremer (1999)), the problem how humans, as finite created beings, can think about the infinite and transcendent God was central to his considerations. The Aristotelian "potential" and "actual" infinities had been transformed into, respectively, "negative" and "positive" infinities. New methods based on the principles of *docta ignorantia* and *coincidentia oppositorum* were being employed. These methods, as Nicholas believed, are applicable to all branches of knowledge, even to the natural sciences. Nicholas developed an original version of Christian Neoplatonism. In *De docta ignorantia* (*On Learned Ignorance*, 1440), a mystical discourse on the finite and the infinite, God is conceived as:

> Maximum itaque absolutum unum est, quod est omnia; in quo omnia, quia maximum. Et quoniam nihil sibi opponitur, secum simul coincidit minimum; quare et in omnibus; et quia absolutum, tunc est actu omne possibile esse, nihil a rebus contrahens, a quo omnia.

> Thus, the Maximum is the Absolute One which is all things. And all things are in the Maximum (for it is the Maximum); and since nothing is opposed to it, the Minimum likewise coincides with it, and hence the Maximum is also in all things. And because it is absolute, it is, actually every possible being; it contracts nothing from things, all of which [derive] from it. (Nicholas of Cusa (1985), 7.5)

God, Absolute Maximum, embraces all things in himself, he unites all opposites: he is the *complicatio omnium contradictoriorum*. The maximum infinite is identified with the minimum infinite.

> In God we must not conceive of distinction and indistinction, for example, as two contradictories, but we must conceive of them as antecedently existing in their own most simple beginning, where distinction is not other than indistinction. (Nicholas of Cusa (1997), p. 29)

Nicholas argued that the human mind needs to realize its own ignorance as to what God is like since there are ontological and cognitive disproportions between God and the finite human knower. Cognition of God is possible only by analogy:

> Omnis igitur inquisitio in comparativa proportione facili vel difficili existit; propter quod infinitum ut infinitum, cum omnem proportionem aufugiat, ignotum est. (Nikolaus von Kues (1989), 5.20)
>
> Therefore, every inquiry proceeds by means of a comparative relation, whether an easy or a difficult one. Hence, the infinite, qua infinite, is unknown; for it escapes all comparative relation. (Nicholas of Cusa (1985), I, 1.3)

To justify the conviction of Cusa concerning of the role mathematics plays in theology, it is enough to cite the titles of chapters 11 and 12 of Nicholas of Cusa (1985) (*On learned ignorance*): 11. Mathematics assists us greatly in apprehending various divine [truths]; and 12. The way in which mathematical signs ought to be used in our undertaking.

The knowledge of mathematics is indispensable to being able to think about God:

> ...ita ut Boethius, ille Romanorum litteratissimus, assereret neminem divinorum scientiam, qui penitus in mathematicis exercitio careret, attingere posse. (Nikolaus von Kues (1989), 23.5)
>
> Thus, Boethius, the most learned of the Romans, affirmed that anyone who altogether lacked skill in mathematics could not attain a knowledge of divine matters. (Nicholas of Cusa (1985), I, 11.31)

To Cusanus, geometry was the best science to use in order to get an understanding of the infinite (Monnoyeur-Broitman (2013)). Let us cite some other chapter titles from Nicholas of Cusa (1985) (*On learned ignorance*): 13. The characteristics of a maximum, infinite line. 14. An infinite line is a triangle. 15. The maximum triangle is a circle and a sphere. 16. In a symbolic way the Maximum is to all things as a maximum line is to [all] lines. 17. Very deep doctrines from the same [symbolism of an infinite line]. 18. From the same [symbolism] we are led to an understanding of the participation in being. 19. The likening of an infinite triangle to maximum trinity. 20. Still more regarding the Trinity. There cannot be fourness, [fiveness], etc., in God. 21. The likening of an infinite circle to oneness.

Cusanus demonstrated how moving beyond the finite toward the infinite in geometry may be by analogy applied to understanding the relation between created beings and their Creator. Cusanus was the first who showed that reasoning about the infinite was not always entirely nonsensical. He paved the way for calculus and a mathematics of the continuum (http://integralscience.org/cusa.html). It seems that Cusanus was that Christian theologian for whom mathematics acquired the highest status in metaphysical deliberations (Drozdek (1995), p. 134). Nicholas of Cusa was much admired by Cantor.

The seventeenth century was an important period in the conceptual development of the notion of the infinite. René Descartes (1596–1650) is a key thinker in the development not only of philosophical thought but also of mathematics and natural science. The modern version of the mind–body problem was one of many of his achievements in formulating a rationalistic and calculative conception of human beings.

To prove the existence of something, we first should know what it is. This proposition is shared also by, e.g., Frege:

> A concept is still admissible even though its defining characteristics do contain a contradiction: all that we are forbidden to do, is to presuppose that something falls under it. But even if a concept contains no contradiction, we still cannot infer that for that reason something falls under it. If such concepts were not admissible, how could we are prove that a concept does not contain any contradiction? It is by no means always obvious; it does not follow that because we see no contradiction there is none there, nor does a clear and full definition afford any guarantee against it. (Frege and Austin (1980), pp. 105–106)

In the case of God the determination of the divine essence is not absolutely adequate and precise. Thus it could not be entirely separated from the demonstration of its existence.

To Descartes only God is absolutely infinite. The Infinite is that which not only does not have limits, but we are also certain that there are no limits. Things for which we do not observe limits and which we cannot prove that they must have no limits or – in other words – that there may or may not still be limits, are merely indefinite (Descartes (1985), 1, xxvii). The infinite is conceived as a positive idea. The indefinite is conceived as negative idea. The infinite could not be constructed from the indefinite, since a positive idea cannot be constructed from the negative idea of the indefinite.

> The point of using 'indefinite' rather than 'infinite' is to reserve 'infinite' for God, because he's the only thing that our understanding positively tells us *doesn't* have any limits. The most we know about anything else is the negative information that *we can't find* any limits in it. (Descartes (2008), 27, p. 7)

Ideas that are innate in the human mind allow perceiving the nature of reality. Due to reflection on our in-born idea of God we know His attributes. Any of these attributes is clearly recognized as infinite perfection. The content of the idea of God as an infinite being is infinite; thus, its cause is also infinite:

> there must be at least as much <reality> in the efficient and total cause as in the effect of that cause. (Descartes (1988), p. 91)

The infinite being is also a perfect being. A perfect being could not be a deceiver:

> It is clear enough from this that he cannot be a deceiver, since it is manifest by the natural light that all fraud and deception depend on some defect. (Descartes (1988), p. 98)

Thus, because we have an idea of an infinite being, God exists. God is conceived as:

> ... the very being the idea of whom is within me, that is, the possessor of all the perfections which I cannot grasp, but can somehow reach in my thought, who is subject to no defects whatsoever. (Descartes (1988), p. 98)

Descartes' argument for the existence of God slightly differs from the famous ontological argument of St. Anselm. Descartes states that:

1. whatever is contained in a clear and distinct idea of a thing must be predicated of that thing;
2. we have the idea of an absolutely perfect Being;
3. a clear and distinct idea of an absolutely perfect Being contains the notion of actual existence.
   Therefore: the absolute Being exists.

Humans are finite hence we are not able fully to recognize the nature of an infinite being. He wrote:

> We will thus never embarrass ourselves by disputes about the infinite, seeing it would be absurd for us who are finite to undertake to determine anything regarding it, and thus as it were to limit it by endeavouring to comprehend it. (Descartes (2015), I, XXVI)

Later Blaise Pascal (1623–1662) would maintain that:

> We know that there is an infinite, and are ignorant of its nature. As we know it to be false that numbers are finite, it is therefore true that there is an infinity in number. But we do not know what it is. It is false that it is even, it is false that it is odd; for the addition of a unit can make no change in its nature. Yet it is a number, and every number is odd or even (this is certainly true of every finite number). So we may well know that there is a God without knowing what He is. …
>
> … We know the existence of the infinite and are ignorant of its nature, because it has extension like us, but not limits like us. But we know neither the existence nor the nature of God, because He has neither extension nor limits.(Pascal (2015), p. 233)

According to Philip Clayton (2000) the development of the concept of God as "infinite" and "perfect" started with Descartes' break with medieval tradition.

John Locke (1632–1704) is one of the most influential thinkers of modern times, initiator of the tradition known as British empiricism. According to him perception and reflection are the only source of all our ideas. He refuted the scholastic model of knowledge and science and rationalistic nativism. Our mind is a *tabula rasa*. He argued that even our ideas of infinity, number, space, substance, and so on, could be acquired by us in our experience. But we do not have a proper idea of the infinite. All sensory data are finite and hence our idea of infinity is merely negative or privative.

In Locke (2015), Book II, ch. XVII, 1, about the infinity of God Locke maintained:

> It is true, that we cannot but be assured, that the great God, of whom and from whom are all things, is incomprehensibly infinite: but yet, when we apply to that first and supreme Being our idea of infinite, in our weak and narrow thoughts, we do it primarily in respect to his duration and ubiquity; and, I think, more figuratively to his power, wisdom, and goodness, and other attributes which are properly inexhaustible and incomprehensible, &c. For, when we call THEM infinite, we have no other idea of this infinity but what carries with it some reflection on, and imitation of, that number or extent of the acts or objects of God's power, wisdom, and goodness, which can never be supposed so great, or so many, which these attributes will not always surmount and exceed, let us multiply them in our thoughts as far as we can, with all the infinity of endless number. I do not pretend to say how these attributes are in God, who is infinitely beyond the reach of our narrow capacities: they do, without doubt, contain in them all possible perfection: but this, I say, is our way of conceiving them, and these our ideas of their infinity.

The philosophical consideration of Baruch Spinoza (1632–1677) that influenced Leibniz and his writings were carefully studied by Cantor. The influence of Spinozean thought onto Cantor's is a subject of consideration by Bussotti and Tapp. They maintain that "the study of Spinoza provides deepening insights into Cantor's philosophical theory, whilst Cantor cannot be called a 'Spinozist' in any stricter sense of that word." (Bussotti and Tapp (2009)). Spinoza's views on the infinite were highly complex and deeply rooted in his view of God and nature. In *Ethics* (*Ethica ordine geometrico demonstrata*) he defined God (http://www.thelatinlibrary.com/spinoza.ethica1.html):

> *Ethica ordine geometrico demonstrata. Pars prima. De Deo. Definitiones* VI. Per Deum intelligo ens absolute infinitum, hoc est, substantiam constantem infinitis attributis, quorum unumquodque aeternam, et infinitam essentiam exprimit.
>
> By God I understand a being absolutely infinite, that is, a substance consisting of an infinity of attributes, of which each one expresses an eternal and infinite essence. (Melamed (2004), p. 211)

A substance is what is in itself and is conceived through itself. An attribute is what the intellect perceives of a substance, as constituting its essence. There is only one substance with infinite attributes. The infinite substance is the only substance. The substance is God; and everything else that is, is in God. Outside of God no other substance is possible. Spinoza reserved a special kind of infinity just for God, one which cannot be defined by all of the finite objects in nature. The infinity of God, the infinity of quantity and infinity of duration are quite different types of infinity. The distinction is based on distinctions among eternity, duration, and time, and the distinction between conceiving a thing by the intellect and conceiving it through imagination. The first kind of infinity is by its own nature and is only conceived by intellect, not by imagination. If imagination is brought into play the second kind is derived from the first. This is mathematical infinity. The third kind is a species of indefiniteness. Something that is absolutely infinite cannot have its infinitude expressed mathematically, and hence is indefinite, though no upper bound can be given in this case. The notion of God cannot be a mathematical notion since it is a notion of the intellect and not of the imagination.

Those who deny the existence of the actual Infinite are criticized. According to Spinoza the denial of actual infinity results from confusion and ignorance.

Nothing can be conceived without God. The Infinite, as the rationalists maintained, is an idea which is presented to the mind immediately, and is not an idea we arrive at by negating the finite. Something is infinite by virtue of its nature and can be rightly called infinite by virtue of its definition. An infinite intellect would comprehend all of the formal essences of things. Human knowledge of the infinite attributes of God is limited to two: extension and thought. The intellect, whether infinite or finite, is only a mode of the attribute of thought. Part II of Ethics is devoted to the "Nature and Origin of the Mind". In the Corollary to Prop. XI it is stated that the human mind is part of the mind of God:

> …mentem humanam partem esse infiniti intellectus Dei ac proinde cum dicimus mentem humanam hoc vel illud percipere, nihil aliud dicimus quam quod Deus non quatenus infinitus est sed quatenus per naturam humanæ mentis explicatur sive quatenus humanæ mentis essentiam constituit, hanc vel illam habet ideam et cum dicimus Deum hanc vel illam ideam habere non tantum quatenus naturam humanæ mentis constituit sed quatenus simul cum mente humana alterius rei etiam habet ideam, tum dicimus mentem humanam rem ex parte sive inadæquate percipere. (http://ethics.spinozism.org/text.php?p=2&lang=la)
>
> …the human mind is part of the infinite intellect of God; thus when we say, that the human mind perceives this or that, we make the assertion, that God has this or that idea, not in so far as he is infinite, but in so far as he is displayed through the nature of the human mind, or in so far as he constitutes the essence of the human mind; and when we say that God has this or that idea, not only in so far as he constitutes the essence of the human mind, but also in so far as he, simultaneously with the human mind, has the further idea of another thing, we assert that the human mind perceives a thing in part or inadequately. (http://www.gutenberg.org/files/3800/3800-h/3800-h.htm#chap02)

Leibniz and Cantor's conceptions of infinity were influenced by Spinoza's philosophy.

Gottfried Wilhelm Leibniz (1646–1716), one of the greatest thinkers of the seventeenth and eighteenth centuries, was convinced that logical principles are innate.[9]

Like Spinoza, Leibniz distinguished three degrees of infinity (Nachtomy (2011)): *Infinitum*, *Maximum*, and *Omnia*. The lowest (*tantum infinitum*) is, e.g., like the asymptote of the hyperbole. Its magnitude is greater than we can expand by sensible things. The second (*maximum in suo silicet genere*) is the whole of space and eternity. The idea of infinity as a whole could be conceived only as an absolute, i.e. as an attribute which has no limits (Leibniz (1982), bk 2, ch. xvii, p. 159). It contains everything of its kind. The highest degree (*Omnia*) is in God. It contains everything. God is an actual infinite, not merely a potential one.

Leibniz rejected the actual infinity in mathematics and accepted it in nature and in his metaphysical system (Friedman (1979), p. 186. See also, Marciszewski (2001)). To Leibniz the world is composed of infinitely many indivisible monads, but only God is absolutely infinite:

> It is perfectly correct to say that there is an infinity of things, i.e. that there are always more of them than one can specify. But it is easy to demonstrate that there is no infinite number, nor any infinite line or other infinite quantity, if these are taken to be genuine wholes. The true infinite, strictly speaking, is only in the absolute, which precedes all composition and is not formed by the addition of parts. (Leibniz (1982), bk 2, ch. xvii, p. 157)

9 Noam Chmosky (1928–) developed the concept that the rules of grammar and ability to use language are innate. This concept has had an impact on advances in programming languages.

Spinoza rejected the claim that nature doesn't contain any actual infinities. Leibniz sought to find a type of infinity within nature. In a letter to Foucher in 1693, he wrote:

> One must be delighted, Sir, that you have given a reasonable sense to the doubts of the Academicians. . . . I will be delighted to one day see their opinions digested and clarified by your efforts.
>
> . . . ; and I now express myself quite differently on the subject of indivisibles, amongst other things. It was the essay of a young man who had still not gone deeply into mathematics. . . . Father Gregory of St. Vincent has very well demonstrated by calculation even of the divisibility to infinity, the place where Achilles must catch the tortoise which is ahead of him, according to the proportion of speeds. Thus geometry serves to dissipate these apparent difficulties.
>
> I am so much in favour of actual infinity that instead of admitting that nature abhors it, as is commonly said, I hold that it assumes it everywhere, in order to better show the perfections of its author. Thus I believe that there is no part of matter that is not, I do not say divisible, but actually divided, and consequently, the least particle must be considered as a world full of an infinity of different creatures. (http://www.leibniz-translations.com/foucher.htm)

Leibniz, like Descartes, makes God's infinity something quite different from mathematical and physical (e.g., time, space) infinities. Nevertheless all the other infinities depend on the theological infinity, which is an attribute of God. On it depends our ability of applying the same rule or process over and over (cf. Le Blanc (1993)). God grasps infinity as a whole. To Him the understanding of infinity is not a result of an analysis of infinite processes. The theological infinite as an attribute of God grounds the mathematical infinite.

To the most insightful mathematical treatises that influenced and anticipated Cantor belongs *Paradoxes of the Infinite* (Paradoxien des Unendlichen) (Bolzano (1851)), a book by Bernard Bolzano (1781–1848), Czech philosopher, theologian and mathematician.

5 Georg Cantor's infinite God

The selection of the above outlined conceptions of God and infinity was guided by their influence on Georg Ferdinand Ludwig Philipp Cantor (1845–1918). Their creators were scholars who – directly or indirectly – affected Cantor's solution of the problems of existence of actual infinity and the infinity of God. They are precursors of Cantor, but without any doubt Cantor founded the conception of the actually infinite, and his solution prevails in contemporary philosophical thought.

Cantor is known as the creator of set theory, the fundamental theory for contemporary mathematics. For Hilbert it was:

> The finest product of mathematical genius and one of the supreme achievements of purely intellectual human activity. (Hilbert (1983), p. 188)

Let us add that some distinguished mathematicians had quite opposite opinions, e.g. for Poincaré it was "grave disease" (Bell (2014), p. 558; Dauben (1979), p. 266; Poincaré (1963)). For Leopold Kronecker, his old professor, Cantor was a "charlatan" and "corrupter of youth". He said:

> I don't know what predominates in Cantor's theory – philosophy or theology, but I am sure that there is no mathematics there. (cf., Tall (2013), p. 363)

Wittgenstein, the philosopher, described Cantor's theory of sets as "laughable nonsense".[10] Perhaps Cantor was the first scholar who really understood the meaning of infinity and gave it mathematical precision. Cantor was aware of such opposition. How was he able to bear all the allegations? Maybe theological motivation gave him enough strength (Neidhart (2008), p. 620). Cantor was undoubtedly the first who realised that there are different kinds, different sizes, of infinity. Joseph Dauben (1992), pp. 59–60, the biographer of Cantor, argues that Cantor's theory of infinity was a revolution (in the sense of Kuhn (1962)) not only in mathematics. Before Cantor infinity was the subject of confused thinking and invalid argumentation. Some self-evident propositions were rejected, e.g. the principle: *Totum est majus sua parte* (the whole is greater than the part).[11] Though Cantor's theory of infinity resolved the old antinomies, it involved new paradoxes. Cantor's theory of sets posed a difficult problem, not for the concept of infinity, but for the fundamental concept of set. Nevertheless, as Hilbert told:

> Aus dem Paradies, das Cantor uns geschaffen, soll uns niemand vertreiben können.
> Out of the paradise that Cantor created for us, no one must be able to expel us. (Hilbert (1925), p. 170)

---

10 For more about Cantor's struggle for the existence of actual infinity, see Dauben (1979, 2004). It is curious that Cantor fiercely opposed infinitesimals, describing them as both an "abomination" and "the cholera bacillus of mathematics".

11 It was already known to Galileo Galilei that in the case of infinite sets it could be that a proper subset of a set is equinumerous with the set. In *Discorsi e Dimostrazioni Matematiche Intorno a Due Nuove Scienze* (published in 1638) (Discourses and Mathematical Demonstrations Relating to Two New Sciences) he discussed what is known as Galileo's Paradox:

Though Wittgenstein replied:

> if one person can see it as a paradise of mathematicians, why should not another see it as a joke? (Wittgenstein (1978), p. 264)

And according to Cantor:

> in the late 1800s, finally created a theory of the actual infinite which by its apparent consistency, demolished the Aristotelian and scholastic "proofs" that no such theory could be found. (cf., Rucker (2007), p. 7)

Cantor himself was surprised by his own discovery that the sets $\Re^n$ are equinumerous, for any $n \in \mathbb{N}$. In 1877 in a letter to Richard Dedekind he wrote:

> Ich sehe es, aber ich glaube es nicht.
> I see it, but I don't believe it!

Due to Cantor's efforts at vindicating the concept of actual infinity, Aristotelian and scholastic tradition seems to have been overcome. This does not mean that Cantor understood actual infinity as they did, in particular as Aristotle did (Le Blanc (1993), p. 52).

Cantor sought precise definitions of the meanings of "infinity"; first of all to determine the most important difference that is between potential and actual infinities (Dewenders (2002), p. 124). He considered various notions of infinity (Décaillot (2011), Anhang 2, Die verschiedene Bedeutungen des Begriffes "unendlich" in der Mathematik). Cantor distinguishes between Uneigentlich-Unendliche (improper-infinity) and Eigentlich-Unendliche (proper-infinity) (Cantor (1932b), p. 165–166). The improper-infinity is in fact a changeable finite with no definite value:

> ...the totality of all numbers is infinite, and that the number of squares is infinite.; neither is the number of squares less than the totality of all numbers, nor the latter greater than the former; and, finally, the attributes "equal", "greater", and "less" are not applicable to the infinite, but only to finite quantities.

The idea that in the case of infinite quantities "greater", "less" and "equal" could not be taken in their "finite" sense was used by R. Dedekind who defined the infinite set as such that is equinumerous with its proper subset. The idea that size can be measured by one-to-one correspondence is known as Hume's principle, although he believed the principle could not be applied to infinite sets.

> The potential infinite means nothing other than an undetermined, variable quantity, always remaining finite, which has to assume values that either become smaller than any finite limit no matter how small, or greater than any finite limit no matter how great. (Cantor (1887))

Contrary to this, the proper-infinity appears in definite form. In the name of potential infinity it would be better to abandon the word "infinity" (Cantor (1932b), p. 404). The Actual-Infinite is not a variable, but rather is fixed and determined in all its parts. It is a genuine constant:

> in truth the potentially infinite has only a borrowed reality, insofar as a potentially infinite concept always points towards a logically prior actually infinite concept whose existence it depends on. (Cantor (1932b), p. 404); quoted in Rucker (2007), p. 3)

The actual infinity holds primacy over every potential infinity. Each potential infinite presupposes the existence of an actual infinity (Hallett (1984), p. 25, and Moore (2001), p. 117). The thesis that potential infinity is ontologically dependent on actual infinity is justified by the fact that potential infinity is a variable:

> In order for there to be a *variable* quantity in some mathematical study, the 'domain' of its variability must strictly speaking be known beforehand through a definition. However, this 'domain' cannot itself be something variable, since otherwise each fixed support for the study would collapse. Thus this domain is a definite, actually infinite set of values. (Hallett (1984), p. 25)

Hence, in Cantor's words:

> ... each potential infinity, if it is rigorous applicable mathematically, presupposes an actual infinite. (Hallett (1984), p. 25: and cf. Cantor (1932a), pp. 224–233)

To Cantor

> There is no doubt that we cannot do without variable quantities in the sense of the potential infinite. But from this very fact the necessity of the actual infinite can be demonstrated. (Cantor (1932a))

The idea of the priority of the infinite over the finite has its source in the thought of Nicholas of Cusa. Descartes rediscovered it:

> I see that there is manifestly more reality in infinite substance than in finite, and therefore that in some way I have in me the notion of the infinite earlier than the finite ... (Descartes (1988), http://www.sacred-texts.com/phi/desc/med.txt)

Cantor argued that finite lines and finite numbers are embedded in the infinite.

The approval of the existence of the actual infinity of mathematical objects distinguishes classical mathematics from various finitistic and intuitionistic mathematics.

Though Cantor's achievements are fundamental for mathematics, his mathematical views were intrinsically linked to their philosophical and theological implications, and he himself, as a religious man, seems more appreciated for his devotion to the theological questions of infinity. Starting his mathematical study, Cantor in a letter to his father declared:

> ...meine Seele, mein ganzes Ich lebt in meinem Berufe; was der Mensch will und kann, und wozu ihn eine unbekannte geheimnisvolle Stimme treibt, *das* führt er durch! (Cantor (1991), p. 19)

His work on the infinite Cantor considered to have been directly communicated to him by God, who had chosen him. In 1894 in a letter to Hermite he wrote:

> But now I thank God, the all-wise and all-good, that He always denied me the fulfillment of this wish [for a position at university in either Göttingen or Berlin] for He thereby constrained me, through a deeper penetration into theology, to serve Him and His Holy Roman Catholic Church[12] better than I been able with my exclusive preoccupation with mathematics. (Dauben (1979), p. 147)

In a letter to Swedish mathematician Mittag-Leffler in 1883 he wrote:

> Ich bin weit davon entfernt, mir meine Entdeckungen zum persönlichen Verdienste anzurechnen, denn ich bin nur ein Werkzeug einer höheren Macht, die nach mir weiter wirken wird, ebenso wie sie vor Jahrtausenden in Euclid und Archimedes sich offenbart hat. (Cantor (1991), p. 160)

Later in 1884 he added:

> ... so ist dies nicht mein Verdienst, ich bin in Bezug auf den Inhalt meiner Arbeiten nur Berichterstatter und Beamter. (Cantor (1991), p. 171)

The assistance of God is also the ultimate reason of reliability of Cantor's set theory. In 1888 Cantor wrote:

12 Let us add that taking into account the religious convictions of Cantor he was in fact Protestant (Cantor (1991), p. 444). Cantor's mother was a Catholic. Moreover it could be also influenced by Aquinas whom he carefully studied. Nevertheless, he continued, and made deeper his connection to the Catholic church. See also, Neidhart (2008), p. 620.

> ...meine Lehre steht felsenfest, jeder gegen sie gerichtete Pfeil wird auf den Schützen selbst zurückschnellen. Woher ich dies weiß? Weil ich sie nach allen Beziehungen seit vielen Jahren erprobt, alle Einwände, die je gegen die unendlichen Zahlen gemacht worden sind, geprüft habe und vor allem, weil ich ihre Wurzeln gewissermaaßen bis zur ersten untrüglichen Ursache alles creatürlichen Seins verfolgt habe. (Cantor (1991), p. 297)
>
> My theory stands as firm as a rock; every arrow directed against it will return quickly to its archer. How do I know this? Because I have studied it from all sides for many years; because I have examined all objections which have ever been made against the infinite numbers; and above all because I have followed its roots, so to speak, to the first infallible cause of all created things. (Dunham (1990), p. 283)

Moreover, to Cantor his theory of infinite is a theology of infinite. In a letter to Thomas Esser (February 1896) Cantor wrote that:

> Von mir wird der christlichen Philosophie zum ersten Mal die wahre Lehre vom Unendlichen in ihren Anfängen dargeboten.
>
> From me, Christian philosophy will be offered for the first time the true theory of the infinite. (Cantor (1991), p. 526)

Some Christian theologians saw Cantor's work as a challenge to their view of the nature of God.

Cantor distinguished three kinds of actual infinity:

> Es wurde das A.-U. [A.-U.: Actual Infnity] nach *drei* Beziehungen unterschieden: *erstens* sofern es in der höchsten Vollkommenheit, im völlig unabhängigen, außweltlichen Sein, in *Deo* realisiert ist, wo ich es *Absolutunendliches* oder Kurzweg *Absolutes* nenne; *zweitens* sofern es in der abhängigen, kreatürlichen Welt vertreten ist; *drittens* sofern es als mathematische Größ, Zahl order Ordnungstypus vom Denken *in abstracto* aufgefaßt werden kann. In den *beiden* letzten Beziehungen, wo es offenbar als beschränktes, noch wieter Vermehrung fähiges und *insofern* den Endlichen *verwandts* A.U. sich darstellt, nenne ich es *Transfinitum* und setze es den *Absoluten* strengsten engegen. (Cantor (1932b), p. 378)

> The actual infinite arises in three contexts: *first* when it is realized in the most complete form, in a fully independent otherworldly being, in Deo, where I call it the Absolute Infinite or simply Absolute; *second* when it occurs in the contingent, created world; *third* when the mind grasps it *in abstracto* as a mathematical magnitude, number or order type. I wish to mark a sharp contrast between the Absolute and what I call the Transfinite, that is, the actual infinities of the last two sorts, which are clearly limited, subject to further increase and thus related to the finite. (Russell (2011), p. 282)

The actual infinites were defined as:

> ...erstens, sofern es *in Deo extramundano aeterno omnipotenti sive natura naturante*, wo es das Absolute heißt, zweitens sofern es *in concreto seu in natura naturata* vorkommt, wo ich es Transfinitum nenne und drittens kann das A.-U. *in abstracto* in Frage gezogen werden, d. h. sofern es von der menschlichen Erkenntnis in Form von aktual-unendlichen, oder wie ich

> sie genannt habe, von transfiniten Zahlen oder in noch allgemeinerer Form der transfiniten Ordnungstypen (αριθμοινοητοι oder ειδητιχοι) aufgefasst werden könne. (Cantor (1932a), p. 372)
>
> ...firstly, inasmuch as it is called in *Deo extramundano aeterno omnipotenti sive natura naturante* ["in God—who is Beyond the World, Eternal, Omnipotent—who gives rise to nature"], where it is called the Absolute, secondly, inasmuch as it occurs *in concreto seu in natura naturata* [or "concretely, in created nature"], where I name it Transfinitum and thirdly the A.-I. can be called into question *in abstracto*, that is inasmuch as it may be comprehended by human cognition [Erkenntnis] in the form of actual-infinite, or as I have named them, transfinite numbers, or in the even more general form of the transfinite ordinal types (αριθμοινοητοι or ειδητιχοι) ["numbers of the mind" or "seen in the eye of the mind"]. (http://www.schillerinstitute.org/fid_91-96/943_transfinite.html#n13)

There were actual infinities: transfinite and absolute:

> Eine *andere* häufige *Verwechselung* geschieht mit beiden Formen des *aktualen* Unedlichen, indem nämlich das *Transfinite* mit dem *Absoluten* vermengt wird, während doch diese Begriffe streng geschieden sind, insofern ersteres ein *zwar Unendliches*, aber doch *noch Vermehbares*, das letztere aber wesentlich als *unvermehrbar* und daher mathematisch *undeterminierbar* zu denken ist; ...(Cantor (1932b), p. 375)

There are two kinds of the notion of "absolute infinity": mathematical and theological (Tapp (2012)). Mathematical infinity is conceived by Cantor as a quantitative concept (not as qualitative). Mathematical absolute infinity differs from the transfinite only in that it is non-augmentable. It is beyond all alephs and ordinal numbers, the size of the proper class of all of them. Maybe Pesch's definition of infinity as: "id, quo non sit maius nec esse possit" (that than which there is nothing bigger or could be) (Pesch (1883), ¶ 403) influenced Cantor to take non-augmentability as the characteristic property of absolute infinity (Tapp (2012), p. 5). The confusion of two forms of the Actual Infinite, namely the Transfinite with the Absolute, takes place in pantheism. The two concepts should be strictly separated, insofar as the Transfinite is to be conceived as an indeed Infinite but nevertheless a yet increasable, and the Absolute, however, essentially as unincreasable and therefore mathematically indeterminable.

There are three levels of existences:

1. *in Intellectu Divino* (in the mind of God);
2. *in abstracto* (in the mind of man); and,
3. *in concreto* (in the physical universe).

The notion of Absolute Infinity was introduced as early as 1882 along with the ordinal theory of cardinality. The Absolute Infinite exists only in the mind of God. Cantor maintains that numbers too, transfinite and cardinal, have existed in the highest level of reality for all eternity as Ideen *in intellectu Divino* (Tapp (2005), Letter to Jeiler, 13.10.1895, p. 427). Inconsistent multiplicities, what we now call proper classes (i.e. classes that taken as sets could be inconsistent), are completed (fertig) only in *intellectu Divino*.

Numbers exist in their intra-subjective reality in the human mind (*in abstracto*) as well as in their trans-subjective reality (*in concreto*). The abstract infinite is found in mathematics. But what about infinity in *natura creata*? The concrete infinite is found in nature. The infinite is identified with Spinoza's "natura naturata" (which inspired controversy about the pantheism of Cantor's views). Following Leibniz, Cantor thought that there were a transfinite number of elementary units: corporeal (matter) and ethereal (ether) monads. Cantor's views are closer to Leibniz than to Spinoza (Newstead (2009)).

In Cantor (1887) we read:

> Das *Transfinite* mit seiner Fülle von Gestaltungen und Gestalten weist mit Nothwendigkeit auf ein *Absolutes* hin, auf das "wahrhaft Unendliche", an dessen Größe keinerlei Hinzufügung oder Abnahme statthaben kann und welches daher quantitativ als *absolutes* Maximum anzusehen ist. Letzteres übersteigt gewissermassen die menschliche Fassungskraft und entzieht sich namentlich mathematischer Determination; wogegen das *Transfinite* nicht nur das weite Gebiet des Möglichen in Gottes Erkenntnis erfüllt, sondern auch ein reiches, stets zunehmendes Feld idealer Forschung darbietet und meiner Überzeugung nach auch in der Welt des Geschaffenen bis zu einem gewissen Grade und in verschiedenen: Beziehungen zur Wirklichkeit und Existenz gelangt, um die Herrlichkeit des Schöpfers, nach dessen absolut freiem Rathschluß, stärker zum Ausdrucke zu bringen, als es durch eine bloß "endliche Welt" hätte geschehen können. Dies wird aber auf allgemeine Anerkennung noch lange zu warten haben, zumal bei den Theologen, so werthvoll auch diese Erkenntnis als Hilfsmittel zur Förderung der von ihnen vertretenen Sache (der Religion) sich erweisen würde. (Cantor (1932b), p. 405)

The meanings of "infinity" are different and should not be confused, though they are – even the "theological" meaning (Neidhart (2008), p. 623) – in mutual close dependence on one another. The distinction between Absolute Infinity and transfinite numbers has a profound impact on our modern worldview.

Cantor was convinced that there are theological consequences of his theory. There is an infinity of transfinite numbers. The term "transfinite" was coined by him, to distinguish various levels of infinite numbers from the Absolute Infinity. The Absolute Infinite transcends any transfinite number and is equated with God. Cantor saw no difference between the Absolute Infinite and God as traditionally conceived. His work on infinity in mathematics was accused of undermining God's infinity, but Cantor argued that God's infinity is the Absolute Infinite, which transcends other forms of infinity (Nagasawa (2011), p. 111). Cardinal Johannes Franzelin wrote:

> I confess however, that in my opinion, that which the author calls the "Transfinitum in natura naturata," can not be defended, and in a certain sense, although the author does not appear to intend it, would contain the error of pantheism.(http://www.schillerinstitute.org/fid_91-96/943_transfinite.html#n13)

Cantor declared that his investigations of the infinite were assisted by God himself (Dauben (2004), pp. 8 and 11–13). In 1888 to Jeiler he wrote:

> I entertain no doubts as to the truths of the transfinites, which I recognized with God's help and which, in their diversity, I have studied for more than twenty years; every year, and almost every day brings me further in this science. (Dauben (1979), p. 147)

For him the potential infinity is a variable and as such it has sense only if a domain of its variability is determined. The domain of a variable in the sense of the potential infinite has to be actual infinite. Thus a potential infinite exists only if an actual infinite exists. The concept of actual infinity could not be acquired from our experience. Everything that could be subject of our experience is finite. The experience could not be source of our concept of something that is actual infinite. The only source of the concept of actual infinity could be something that is actually infinite. Actual infinity transcends our finite understanding. Thus if we have the concept of actual infinity, it is given to us by an infinite being. God is an absolute infinite being.

Cantor argued that even God is unable to have direct knowledge about infinity. All infinity is in some ineffable way made finite to God, for it is comprehended by his knowledge. It is what was stated by St. Augustine in book 12, chapter 18 *Against Those Who Assert that Things that are Infinite Cannot Be Comprehended by the Knowledge of God* in *City of God*:

> Far be it, then, from us to doubt that all number is known to Him "whose understanding," according to the Psalmist, "is infinite." The infinity of number, though there be no numbering of infinite numbers, is yet not incomprehensible by Him whose understanding is infinite. And thus, if everything which is comprehended is defined or made finite by the comprehension of him who knows it, then all infinity is in some ineffable way made finite to God, for it is comprehensible by His knowledge. (Schaff (1890), Book 12, chapter 18, pp. 345–346)

6 Mathematics and analogical knowledge about infinite God

Each of the discussed concepts of divine infinity is essentially dependent on the theoretical context, on the philosophy in which it is used. To any of the scholars, in any of the philosophical systems, the infinity of God is conceived in various ways. There are some resemblances – mainly due to their aim to explain Biblical concept(s) of infinity, but there are also deep differences, as e.g. between St. Thomas Aquinas and Descartes and even deeper in the philosophy of Spinoza. But besides being related to the Bible, they are also – which is important from the philosophical angle – related to mathematics. This is the solid fundament of the discussion and comparison. Due to this fact the consideration of infinity can lead and does lead to mature fruit for mathematics, metaphysics, and theology.

Theology should be independent of science, both natural and formal – and vice versa. There should be no place of conflict between both kinds of knowledge. To Max Planck (1858–1947), for example, religion and science are compatible and there is complete concordance:

> Science deals with the objective, material world..... Religion, on the other hand, deals with the world of values. ...In science we are concerned to discover what is true or false; in religion with what is good or evil, noble or base. Science is the basis of technology, religion the basis of ethics. In short, the conflict between the two, which has been raging since the eighteenth century, seems founded on a misunderstanding, or, more precisely, on a confusion of the images and parables of religion with scientific statements. (Planck (1948), Planck (1948), https://edge.org/conversation/science-and-religion)

The independence of theology and science each from the other, does not mean that there are excluded some inspirations of one by the other or any other associations. For example, to Georg Cantor – as a religious man – the theological acceptance of his set theory was very important, but it does not mean that mathematicians who develop set theory are believers in God. Max Planck stated that:

> Both Religion and science require a belief in God. For believers, God is in the beginning, and for physicists He is at the end of all considerations ...To the former He is the foundation, to the latter, the crown of the edifice of every generalized world view. (Planck (1958))

Let us add an opinion of Werner Heisenberg:

> In the history of science, ever since the famous trial of Galileo, it has repeatedly been claimed that scientific truth cannot be reconciled with the religious interpretation of the world. Although I am now convinced that scientific truth is unassailable in its own field, I have never found it possible to dismiss the content of religious thinking as simply part of an outmoded phase in the consciousness of mankind, a part we shall have to give up from now on. Thus in the course of my life I have repeatedly been compelled to ponder on the relationship of these two regions of thought, for I have never been able to doubt the reality of that to which they point. (Heisenberg (1974), p. 213)

Scientific theses should not depend on the theses of theology and vice versa. The postulate of the independence of branches of knowledge is satisfied only if the sets of characteristic theoretical notions of these branches are disjoint.

The theological notion of infinity is not the same as the mathematical or physical. It is neither an infinite quantity nor an infinite magnitude. The acceptance of the thesis of the infinity of God does not have any logical implications for concepts of infinity in mathematics or in natural science. From purely ontological reasons it is possible to accept the existence of actually infinite mathematical objects and to reject the existence of divine infinity. Those who reject the existence of actually infinite mathematical objects can without any inconsistency accept the existence of actual infinity as it is conceived in theology.

The particular sense of divine infinity is stressed by Cantor:

> ...gebrauche ich das Wort "absolut" nur für das, was *nicht mehr vergrösert*, resp. verwolkommnet werde kann, in Analogie des "Absoluten" in den Metaphysik. Meine eigentlich unendlich oder, wenn Sie lieber wollen, transfinite Zahlen w, w+1,...sind nicht "absolut", wie sie, obgleich nicht endlich, dennoch der Vergröß fähig sind. Das Absolute ist jednoch keine Vergrößrung fähig und daher auch für uns inaccessible. (Cantor (1932b), p. 138)

Infinite concepts cannot be acquired by finite means and on the basis of finite experiences[13]: all things we have been capable of observing and measuring are finite. Any definition is finite, i.e. done with the use of a finite number of known concepts. Infinite definitions (which are not done in a finite time) are absurd:

> Eine *jede Definition* ist aber ihrem Wesen nach eine *endliche*, d.h., sie erklärt den zu bestimmenden Begriff durch eine *endliche* Anzahl *bereits bekannter* Begriffe $B_1, B_2, B_3, \ldots, B_n$. "Unendliche Definitionen" (die nicht in endlicher Zeit verlaufen) sind *Undinge*. (Cantor (1991), p. 446)

By definition only finite concepts can be acquired. For instance, the human mind can embrace (at most) any rational number (a pair of natural numbers). There are $\aleph_0$ rational numbers. Some real numbers are not rational. In such a case the name of the real number is an infinite sequence of digits and as such cannot be embraced by a human mind. Some real numbers can be described finitely by means of an algorithm, but there are such numbers for which it cannot be done (Chaitin (1966, 1987, 1997, 2004) and Trzęsicki (2006)). The set of finitely definable real numbers is a countable infinite. But, which is proved by Cantor with the famous diagonal method, the set of real numbers is not a countable infinite (Cantor (1874)). If real numbers which are not finitely described exist, then real

13 The question of acquiring of concepts of infinity is the subject of my paper Trzęsicki (2015).

numbers are not a human mind-dependent entity. Thus the human mental experience could not be the source of acquiring the actual infinity (of real numbers).

According to Cantor, God instilled the concept of number, both finite and transfinite, into the mind of man:

> sowohl getrennt als auch in ihrer aktual unendlichen Totalität als ewige Ideen *in intellectu Divino* im höchsten Grad der Realität existieren. (Cantor (1991), pp. 275f)

This means that transfinite numbers exist in the mind of man, as eternal ideas exist in the mind of God (Dauben (1979), pp. 228–232). God put them into man's mind to reflect his own perfection (Dauben (1979), p. 126). Infinite concepts are innate. To Nicholas of Cusa the Absolute Infinite remains at the border of comprehensibility.

> The Absolute can only be acknowledged and admitted, never known, not even approximately. (Hallett (1984), p. 13)

To divine operation "our human being expands beyond its own boundaries into the infinite" (Staniloae (1994), p. 144).

The only way to acquire a concept of infinity is through its endowment by God. Only to Him do we owe it. This is true for infinity in any sense, theological, mathematical, or physical. This single possibility is clearly stressed by Descartes. If so, if the concept of infinity has its source in God, then it can be used to justify charging with theology those mathematicians who apply concepts and arguments involved in infinite mathematics. Nevertheless, even if the concepts of infinity have their source in a mind illuminated by God, the mathematical and physical concepts, in what they say positively, are independent of the theological concept. Dauben states:

> Later generations might dismiss the philosophy, look askance at his abundant references to St. Thomas or to the Church Fathers, overlook the metaphysical pronouncements and miss entirely the deeply religious roots of Cantor's later faith in the absolute truth of his theory. But all these commitments contributed to Cantor's resolve not to abandon the transfinite numbers. Opposition seems to have strengthened his determination. His forbearance, as much as anything else he might have contributed, ensured that set theory would survive the early years of doubt and denunciation to flourish eventually as a vigorous, revolutionary force in 20th-century mathematics. (Dauben (2004))

We are not able to have any positive knowledge about God, which is in accordance with Christian theology: *si comprehendis non est Deus*. The Christian God is infinite not as a being that has no known limits or bounds, but as a being that has no limits or bounds. The Infinity of God means that He exceeds all the bounds which confine us. Thus our knowledge of Him can be only negative.

Let us ask how the negative knowledge about the infinite God can be acquired by us in a methodological way. The above discussion justifies the answer: by infinite mathematics. Infinity itself is not God, but God is essentially infinite. Mathematicians have systematized and rationalized the treatment of actual infinity in mathematics. My chief claim is that the mathematics of infinity be fruitful in religious thought and also that religious thought of infinity would be still as fruitful as it has been for the development of mathematics.

The infinite not only was inspired by the theological problem of the infinity of God, but also is the best tool to advance understanding of the infinity of God. The analogy between divine and mathematical infinities is quite fruitful and "brings into light the experience of God as the mysterious unspeakable" (https://math.dartmouth.edu/~matc/Readers/HowManyAngels/Blanc.html). Even if the mathematical infinite does not suffice to describe or explain God's infiniteness, it can symbolize it or at least supply us with a useful metaphor: many of the assertions we make about God and about a mathematical infinity are similar. Mathematicians come closer to understanding infinity than theologians. There are worked out principles and rules for reasoning about infinite mathematical problems. Reasoning about the infinite requires abandoning many of the rules that apply to the finite. Though the term "infinite" is equivocal when used theologically, we arrive at a better understanding of God through the language of mathematics.

The knowledge of infinite God is itself infinite. We as finite beings are not able to exhaust the truth about Him. Our knowledge of infinite mathematical domains remains – which Gödel's theorems on undecidability and incompleteness (Gödel (1931, 1934)) imply – potentially infinite. The same is true about our knowledge of God. Let us consider some more examples of using mathematics to theological questions.

There is no contradiction in that there are many kinds of actual infinities. Thus there is no contradiction in that the Holy Trinity is composed of three different infinite beings. Piotr Łukowski (2011) considers Jan Łukasiewicz's (Łukasiewicz (1910), pp. 35–36) comments to the question of Holy Trinity in the Athanasian Creed, which "stands alone in its detailed and beautiful description of the Holy Trinity." (http://www.beginningcatholic.com/athanasian-creed.html). Łukowski cites the point 256 of the *Catechism of Catholic Church* in which is a quotation from St. Gregory Nazianzen's *Orationes*:

> ... give you but one divinity and power, existing one in three, and containing the three in a distinct way. Divinity without disparity of substance or nature, without superior degree that raises up or inferior that casts down ... the infinite co-naturally of three infinities. Each

> person considered in himself is entirely God ... the three considered together ... (Łukowski (2011), p. 30)

Łukowski remarks that:

> A simple division of an infinite set into three sets equipollent with the original one is not sufficient to construct the model, since each person of the Trinity is God whole. This means that every subset should include the whole set.

After rejecting a solution based on St. Gregory Nazianzen, Łukowski (2011), p. 31, proposes a model consisting of three infinite sequences that, though they are completely different, are in some sense one and the same sequence; moreover each of the three sequences is a proper sub-sequence of every other one. According to him it is the right model of the Holy Trinity as it is described in the *Catechism of Catholic Church.*

Mathematicians use the phrase "all but a finite number of members". Any infinite set after the ablating of finite elements remains infinite. This model was already applied by Boethius and can be used in the theology of Providence. The infinite essence of God is the inexhaustible source from which all other beings of the finite world can draw everything they desire to possess. The same is true about salvation: No finite sacrifice is enough, thus only atonement by the suffering of the infinite redeemer could be a potentially infinite source of forgiveness committed by a potentially infinite number of people for their sins against the Infinite.

The Bible appreciates the force of the principle that the Infinite and the finite do not compare. Blaise Pascal (1623–1662), French mathematician, physicist, inventor, writer, and Christian philosopher, uses the argument known already to Boethius that the actually infinite bears no relation to the finite, no matter how large the latter is: *inter finiti et infinite non est proportio*. In Locke's words:

> What I say of man, I say of all finite beings; who, though they may far exceed man in knowledge and power, yet are no more than the meanest creature, in comparison with God himself. Finite or any magnitude holds not any proportion to infinite. (Locke (2015), Book II, ch. XV, 12)

According to Christianity human beings are important as having the grace of being sons of God. But our "infinitesimal" share in the ration should keep us appropriately humble.

7 Conclusions

In the discussed conceptions of the infinity of God it has been stated that our knowledge of God could be only negative. The sense of the phrase "negative knowledge" was pointed out by examples. It could be associated with the negative form of propositions. If so what about double negated propositions? Can we with the help of Cantor's set theory give a better description of the concept of negative knowledge? Positive knowledge about x is such that there is a logical function $f(x)$; the class of values of it is the set $\{x : f(x)\}$. The knowledge is negative only if it is not possible. In his famous antinomy of the set of all sets, Russell proved the class of all sets is not a set; thus there is no function f. Our knowledge about the class of all sets could not be positive. In the proposed sense of "negative knowledge", let us name it "absolute", the knowledge expressed by a negative proposition not necessary is absolute negative. It remains true that God possesses every perfection in its complete and absolute and inexhaustible fullness and – as Nicholas of Cusa closes his treatise on learned ignorance:

> From these [observations] it is clear (1) that in theological matters negations are true and affirmations are inadequate, and (2) that, nonetheless, the negations which remove the more imperfect things from the most Perfect are truer than the others. ... Therefrom we conclude that the precise truth shines incomprehensibly within the darkness of our ignorance. (Nicholas of Cusa (1985), I, 26, 89)

Nevertheless we are still trying to know more and mathematics still seems to be the best tool to know more about the infinity of God. There are still new attempts to use mathematical concepts and tools to know what is not knowable (Steinhart (2009) and Kleszcz (2012)).

Acknowledgement

My thanks go to Professor Witold Marciszewski for his vivid interest in this research as expressed in our friendly discussions.

Bibliography

Aristotle (1930), *Physics*, Vol. 2, edited by W. Ross, translated by R. P. Hardie and R. K. Gaye, Chicago: Clarendon Press.

Augustine of Hippo (1887), *The City of God*, Vol. 2, edited by P. Schaff, Buffalo, NY: Christian Literature Publishing Co. Retrieved from http://www.newadvent.org/fathers/1201.htm

Bell, E. T. (2014), *Men of Mathematics. The Lives and Achievements of the Great Mathematicians from Zeno to Poincaré*, New York: Simo & Schuster.

Boethius (2007), *The Consolation of Philosophy of Boethius*, translated by H. R. James, eBooks & Adelaide. Retrieved from http://ebooks.adelaide.edu.au/b/boethius/ consolation

Bolzano, B. (1851), *Paradoxien des Unendlichen* (2nd ed), Leipzig. Retrieved from https://archive.org/details/paradoxiendesune00bolz

Bolzano, B. (1950), *Paradoxes of the infinite*, London: Routledge and Paul.

Bolzano, B. (1975), *Paradoxien des Unendlichen*, Vol. 99, edited by B. von Rootselaar, Hamburg: F. Meiner.

Bussotti, P. and Tapp, Ch. (2009), "The Influence of Spinoza's Concept of Infinity on Cantor's Set Theory", in *Studies in History and Philosophy of Science*: 40, 25–35.

Cantor, G. (1874), "Über eine Eigenschaft des Inbegriffes alle reellen algebraischen Zahlen", in *Journal für die reine und angewandte Mathematik*: 70, 258–262. (Also, in Cantor (1932b), 115–118.)

Cantor, G. (1883), *Grundlagen einer allgemeinen Mannigfaltigkeitslehre*: *Ein mathematisch-philosophischer Versuch in der Lehre des Unendlichen*, Leipzig. (Also, in Cantor (1932b), 165–208.)

Cantor, G. (1887), "Mitteilungen zur Lehre vom Transfiniten", *Zeitschrift für Philosophie und philosophische Kritik*: 9, 81–125 and 252–270. (Also, in Cantor (1932b), 378–439.)

Cantor, G. (1932a), "Beiträge zur Begründung der transfiniten Mengenlehre", in Cantor (1932b), 282–355. Berlin/Göttingen: Springer.

Cantor, G. (1932b), *Gesammelte Abhandlungen mathematischen und philosophischen Inhalts*, edited by E. Zermelo, Berlin/Göttingen: Springer.

Cantor, G. (1932c), "Mitteilungen zur Lehre vom Transfiniten", in Cantor (1932b), 378–439. Berlin/Göttingen: Springer.

Cantor, G. (1991), *Georg Cantor*: *Briefe*, edited by H. Meschkowski and W. Nilson, Berlin: Springer.

Chaitin, G. J. (1966), "On the Length of Programs for Computing Finite Binary Sequences", in *Journal of the Association of Computing Machinery*: 13, 547–569.

Chaitin, G. J. (1987), *Algorithmic Information Theory*, Cambridge: Cambridge University Press.

Chaitin, G. J. (1997), "An Invitation to Algorithmic Information Theory", in *Dmtcs'96 Proceedings*, Singapore: Springer-Verlag, 1–23.

Chaitin, G. J. (2004), *Leibniz, Randomness & the Halting Probability*. Retrieved from http://www.cs.auckland.ac.nz/CDMTCS/chaitin/ turing.html

Clayton, P. (2000), *The problem of God in modern thought*, Cambridge: William B. Eerdmans Publishing Co.

Dauben, J. W. (1979), *His Mathematics and Philosophy of the Infinite*, Cambridge: Harvard University Press. Retrieved from http://www.math.dartmouth.edu/ matc/Readers/ HowManyAngels/Cantor/Cantor.html

Dauben, J. W. (1992), "Conceptual Revolutions and History of Mathematics: Two Studies in the Growth of Knowledge", in *Revolutions in Mathematics*, edited by D. Gilles, Oxford: Claredon Press, 49–71.

Dauben, J. W. (2004), "Georg Cantor and the Battle for Transfinite Set Theory", in *In Proceedings of the 9th ACMS Conference*, Santa Barbara, CA: Westmont College., 1-22. Retrieved from http://www.acmsonline .org/journal/2004/jacms2004.html (Much of the information presented here may be found in Dauben (1979). Additional details depend on archival material drawn from archives in Gottingen and Berlin, including the Deutsche Akademie der Wissenschaften (Berlin), the Deutsches Zentralarchiv in Merseburg, and the important holdings at the Institute Mittag-Leffler in Djursholm, Sweden.)

Davis, M. (1965), *The Undecidable: Basic Papers on Undecidable Propositions, Unsolvable Poblems, and Computable Functions*, Hewlett, N.Y.: Raven Press.

Décaillot, A. -M. (2008), *Cantor et al France. Correspondance du mathématicien alleman avec les français á la fin du xixe sciècle*, Paris: Éditions Kimé.

Décaillot, A. -M. (2011), *Cantor und die Franzosen: Mathematik, Philosophie und das Unendliche*, translated by Klaus Volkert (?), Berlin, Heidelberg: Springer.

Descartes, R. (1985), "Principles of Philosophy", in *The Philosophical Writings of Descartes*, Vol. 1, translated by J. Cottingham, R. Stoothoff and D. Murdoch, Cambridge: Cambridge University Press, 177–292. Retrieved from http:// ebooks.cambridge.org/chapter.jsf?bid=CBO9780511805042cid= CBO9780511805042A020tabName=Chapter.

Descartes, R. (1988), "Meditations on First Philosophy", in *Descartes: Selected Philosophical Writnigs*, translated by J. Cottingham, R. Stoothoff and D. Murdoch; with an introduction by J. Cottingham, Cambridge: Cambridge University Press, 73–122.

Descartes, R. (2008), *Principles of Philosophy*, edited by J. Bennett. Retrieved from https://www.google.pl/search?q=Jonathan+Bennett+ principles+of+philosophyie=utf-8oe=utf-8gws$_r d = crei = a6PIVdjpJ8HsUOW - uMAM$.

Descartes, R. (2015), *Principles of Philosophy*, BookRix. Retrieved from https://books.google.pl/books?id=GMWzAwAAQBAJ.

Dewender, T. (2002), *Das Problem des Unendlichen im ausgehenden 14. Jahrhundert: Eine Studie mit Textedition zum Physikkommentare des Lorenz von Lindores*, Vol. 36, Amsterdam: B. R. Grüner Publishing Company.

Drozdek, A. (1995), "Beyond Infinity: Augustine and Cantor", in *Laval théologique et philosophique*: 51(1), 127–140. Retrieved from www.erudit .org/revue/ltp/1995/v51/n1/400897ar.pdf

Dunham, W. (1990), *Journey Through Genius. The Great Theorems of mathematics*, New York: John Wiley & Sons, Inc..

Frege, G. and Austin, J. (1980), *The Fundations of Arithmetic: A Logicomathematical Equiry into the Concept of Number*, Northwestern University Press. Retrieved from http://books.google.pl/books?id= z0KtOtNYMEQC

Friedman, J. I. (1979), "On Some Relations Between Leibniz' Monadology and Transfinite Set Theory: A Complement to Russell's Thesis on Leibniz", in *Bertrand Russell Memorial Volume*, Vol. 95, edited by G. W. Roberts, Routledge, 182–221.

Gödel, K. (1931), "Über formal unentscheidbare Sätze der Principia Mathematica und verwandter Systeme I", in *Monatshefte für Mathematik und Physik*: 38, 173–198. Translated in: Gödel (1986), pp. 144–195 – includes the German text and parallel English translation. English translations in: Davis (1965), pp. 5– 38), and van Heijenoort (1967), pp. 595–616. The online version (translation by B. Meltzer, 1962): http://www.ddc.net/ygg/etext/godel/, PDF; and the online version (translation by M. Hirzel): http://nago.cs.colorado.edu/ hirzel/papers/canon00-goedel.pdf.

Gödel, K. (1934), *On Undecidable Propositions of Formal Mthematical Systems*, (mimeographed lecture notes, taken by S. C. Kleene and J. B. Rosser at the Institute for Advanced Study), Princeton. Reprinted in Davis (1965), 39–74.

Gödel, K. (1986), *Collected Works. Publications 1929–1936*, Vil. 1, edited by S. Feferman, J. W. Dawson Jr., S. C. Kleene, G. H. Moore, R. M. Solovay, and J. van Heijenoort, Oxford: Oxford University Press.

Hallett, M. (1984), *Cantorian Set Theory and Limitation of Size*, Oxford: Clarendon Press.

Heisenberg, W. (1974), *Across the Frontiers*, New York: Harper & Row.

Hilbert, D. (1899), *Grundlagen der Geometrie*, Leipzig: Teubner-Verlag. (8th amended ed. by P. Bernays with supplements, Stuttgart, 1956—1987; 14th ed., with contributions by H. Kiechle et al., M. Toepell, ed., Stuttgart, 1999; the first English translation The Foundations of Geometry by E. J. Townsend, Open Court, Chicago, 1902; new translation by L. Unger, Open Court, La Salle, 1972. [The first German edition of the Grundlagen was a contribution to the Festschrift zur Enthüllung des Gauss-Weber Denkmals in Göttingen, pp. 1–92; later editions were published separately]).

Hilbert, D. (1925), "Über das Unendliche", in *Mathematische Annalen*: 95, 161–190. (Lecture given in Münster, 4 June 1925. Also in Hilbert (1899), 7th ed., pp. 262–288, shortened version in Jahresber. Deutsch. Math. Verein. 36 (1927), 201–215; English translation (by S. Bauer- Mengelberg) in (van Heijenoort, 1967), pp. 367–92; 2nd ed., 1971; French translation (by A.Weil), Sur l'infinie, Acta Math. 48 (1926) 91– 122).

Hilbert, D. (1983), "On the Infinite", in *Philosophy of Mathematics. Selected Readings*, edited by P. Benacerraf and H. Putnam, Oxford: Oxford Univeristy Press, 183–201.

Jaspers, K. (1964), *Nikolaus Cusanus*, München: Piper & Co. Verlag.

Kleszcz, R. (2012), "Logika, Wszechmoc, Bóg", in *Filo-Sofija*: 4(19), 37–52.

Knorr, W. R. (1982), "Infinity and Continuity: The Interaction of Mathematics and Philosophy in Antiquity", in *Infinity and Continuity in Ancient and Medieval Thought*, edited by N. Kretzmann, Ithaca and London: Cornell University Press, 112–145.

Kremer, K. (1999), *Nikolaus von Kues* (1401-1464): *einer der grossten Deutschen des 15. Jahrhunderts*, Trier: Paulinus.

Kuhn, T. S. (1962), *The Structure of Scientific Revolutions*, Chicago: University of Chicago Press.

Le Blanc, J. (1993), "Infinity in Theology and Mathematics", in *Religious Studies*: 29(1), 51–62. Retrieved from https://math.dartmouth.edu/ matc/Readers/HowManyAngels/Blanc.html.

Leibniz, G. W. (1982), *New Essays on Human Understanding*, Cambridge: Cambridge University Press.

Locke, J. (2015), *An Essay Concerning Human Understanding*, The University of Adelaide Library. Retrieved from http://www.gutenberg.org/ cache/epub/10615/pg10615.html.

Logan, I. (2009), *Reading Anselm's Proslogion. The History of Anselm's Argument and its Significance Today*, Burlington: Asgate.

Łukasiewicz, J. (1910), *O Zasadzie Sprzeczności u Arystotelesa*, Kraków: AU.

Łukowski, P. (2011), *Paradoxes*, Dordrecht: Springer.

Maor, E. (1987), *To Infinity and Beyond: A Cultural History of the Infinite*, Boston, MA: Birkhäuser. (Reprinted: Princeton University Press, 1991).

Marciszewski, W. (2001), "Leibniz's Mathematical and Philosophical Approaches to Actual Infinity. A Case of Cultural Resistance", in *Studies in Logic, Grammar and Rhetoric*): 4(17), 7–19.

Melamed, Y. (2004), "Hasdai Crescas and Spinoza on Actual Infinity and the Infinity of God's Attributes", in *Sponoza and Medieval Jewish Philosophy*), edited by S. Nadler), Cambridge: Cambridge University Press), 204–215.

Monnoyeur-Broitman, F. (2013), "Nicholas of Cusa's Methodology of the Infinite", in *History & Philosophy of Infinity*), Cambridge: University of Cambridge). Retrieved from http://www.researchgate.net/ profile/Francoise$_M$onnoyeur2/publicationsdoi : DOI : 10.13140/RG.2.1.1595.0881.

Moore, A. W. (2001), *The Infinite*, New York: Routledge).

Nachtomy, O.(2011), "A Tale of Two Thinkers, One Meeting and Three Degrees of Infinity: Leibniz and Spinoza 1675–8", in *British Journal for the History of Philosophy*: 19(5), 935–961. Retrieved from http:// dx.doi.org/10.1080/09608788.2011.599570.

Nagasawa, Y. (2011), *The Existence of God. A Philosophical Introduction*, New York: Routlege.

Neidhart, L. (2008), *Unendlichkeit im Schnittpunkt von Mathematik und Theologie*, Göttingen: Cuvillier Verlag. (Erste Auflage 2007, zweite, überarbeitete Auflage 2008).

Newman, J. R. (1956), *The World of Mathematics*, New York: Simon and Schuster. (Reprinted in 2003, New York: Dover).

Newstead, A. (2009), "Cantor on Infinity in Nature, Numbers, and the Divine Mind", in American Catholic Philosophical Querterly: 83(4), 533– 553. Retrieved from http://www.academia.edu/188281/Cantor$_o$n$_I$nfinity$_i$n$_N$ature$_N$umber$_a$nd$_t$he$_D$ivine$_M$ind.

Nicholas of Cusa (1985), *On Learned Ignorance* (de docta ignorantia), (2nd ed., Vol. 1), edited by P. Wilpert, Minneapolis, Minnesota: The Arthur J. Banning Press. Retrieved from www1.umn.edu/ships/galileo/library/ cusa2.pdf (The translation of Book I was made from De docta ignorantia. Die belehrte Unwissenheit, Book I (Hamburg: Felix Meiner, 1970, 2nd edition), text edited by Paul Wilpert, revised by Hans G. Senger.)

Nicholas of Cusa (1997), *Selected Spiritual Writings*, edited by ...., New York: Paulist Press.

Nikolaus von Kues (1989), *Philosophisch-theologische Schriften*, herausgegeben und eingeführt von Leo Gabriel, Wien: Verlag Herder. Retrieved from http://urts99.uni-trier.de/cusanus/ content/uebs.php?ueb=1 (Übersetzt und kommentiert von Dietlind und Wilhelm Dupré).

Pascal, B. (2015), *Penses: Thoughts on Religion*, translated by W.F. Trotte, Brookfield, Il: Letcetera, Ltd..

Pesch, T. (1883), *Institutiones philosophiae naturalis: Secundum principia s. Thomas Aquinatis ad usum scholasticum*, Freiburg: Herder.

Planck, M. (1948), *Wissenschaftliche Selbstbiographie. Mit einem Bildnis und der von Max von Laue gehaltenen Traueransprache*, Leipzig: Johann Ambrosius Barth Verlag.

Planck, M. (1949), *Scientific Autobiography and other Papers*, translated by F. Gaynor, New York: ....

Planck, M. (1958), *Religion und Naturwissenschaft*, Leipzig: Johann Ambrosius Barth Verlag.

Poincaré, H. (1963), *Mathematics and Science: Last Essays*, New York: Dover Publications.

Rucker, R. (2007), *Infinity and the Mind*, 3rd ed., New Delhi: New Age International, Publishers. (1st ed.: Princeton University Press, 1995).

Russell, R. J. (2011), "God and Infinity: Theological Insights from Cantor's Mathematics", in *Infinity. New Research Frontiers*, edited by M. Heller and W. H.Woodin, Cambridge: Cambridge University Press, 275–289.

Schaff, P. (1890), *St. Augustine's City of God and Christian Doctrine*, New York: Grand Rapids, MI. Retrieved from http://www.ccel.org/ ccel/schaff/npnf102.html.

Simonyi, K. (2012), *A Cultural History of Physics*, A. K. Peters/CRC Press. Retrieved from http://edge.org/conversation/a-cultural-history -of-physics (Excerpt translated by David Kramer).

Sinnige, T. G. (1968), *Matter and Infinity in the Presocratic Schools and Plato*, Vol. 17, Assen: Van Gorcum and Co.. Retrieved from https://books.google.pl/books?id=8pmbtwAACAAJ.

Staniloae, D. (1994), *Orthodox Dogmatics Theology. The Experience of God*, Brookline, Mass.: Holy Cross Orthodox Press.

Steinhart, E. (2009), "A Mathematical Model of Divine Infinity", in *Theology and Science*: 7(3), 261–274.

Sweeney, L. (2012), *Infinity in the Presocratics: A Bibliographical and Philosophical Study*, Springer Science & Business Media.

Tall, D. (2013), *How Humans Learn to Think Mathematically: Exploring the Three Worlds of Mathematics*, Cambridge: Cambridge University Press.

Tapp, Ch. (2005), *Kardinalität und Kardinäle: Wissenschaftshistorische Aufarbeitung der Korrespondenz zwischen Georg Cantor und katholischen Theologen seiner Zeit*, Wiesbaden: Franz Steiner Verlag.

Tapp, Ch. (2012), *Absolute Infinity – A Bridge between Mathematics and Theology?* Wiesbaden: Franz Steiner Verlag. Retrieved from http://foundationaladventures.files .wordpress.com/2012/01/tapp.pdf (An earlier version of this paper was presented during the conference honoring Harvey Friedman, May 15, 2009. Retrieved 03.08.2014).

Tapp, Ch. (2015), "Infinity in Aquinas' Doctrine of God", in *Analytically Oriented Thomism*, edited by M. Szatkowski, Editiones Scholasticae.

Thomas Aquinas (1920), *Summa Theologica*, Online Edition Copyright c 2008 by Kevin Knight. Retrieved from http://www.newadvent.org/summa/index.html (Second and Revised Edition. Literally translated by Fathers of the English Dominican Province).

Thomas Aquinas (1947), *Summa Theologica*, Benziger Bros. Retrieved from http://dhspriory.org/thomas/summa/FP.html (Translated by Fathers of the English Dominican Province).

Thomas Aquinas (1955–1957), *Contra Gentiles*, New York: Hanover House. Retrieved from http://dhspriory.org/thomas/ContraGentiles.htm (Edited, with English, especially Scriptural references, updated by Joseph Kenny, O.P.).

Torretti, R. (1978), *Philosophy of Geometry from Rieman to Poincaré*, Dordrecht: D. Reidel Publishing Company.

Trzęsicki, K. (2006), "From the Idea of Decidability to the Number W", in Studies in Grammar, Logic and Rethoric: 9(22), 73–142. Retrieved from http:// logika.uwb.edu.pl/studies.

Trzęsicki, K. (2015), "How are Concepts of Infinity Acquired?", in Studies in Grammar, Logic and Rethoric: 40(53), 179–217.

van Heijenoort, J. (ed.) (1967), *From Frege to Gödel. A Source Book in Mthematical Logic* 1879—1931, Cambridge Mass.: Harvard University Press.

Wittgenstein, L. (1978), *Remarks on the Foundations of Mthematics*, edited by R. R. G. H. von Wright and G. E. M. Anscombe, translated by G. E. M. Anscombe, Basil Blackwell.

Authors of Contributed Papers

Andrea C. Bottani is Professor of Theoretical Philosophy at the University of Bergamo (Italy) and invited Professor at the University 'San Raffaele' of Milan and at the FTL of Lugano (CH). He taught and did research at the Universities of Genova, Fribourg, Neuchâtel and Urbino. He was Senior Research Fellow at King's College London, Visiting Scholar at the Columbia University New York, President of the Italian Society for Analytic Philosophy, Member of the Steering Committee of the Italian Society for Logic and Philosophy of Sciences and a founder of the Italian Society for Theoretical Philosophy. He is currently member of the Board of Teachers of the Doctorate School "F.I.N.O." (Northwestern Italian Philosophy Consortium) in Philosophy. Among his publications are two authored books, nine edited volumes and more than sixty book chapters and journal articles. His area of specialization includes metaphysics, ontology and philosophy of language.

Paul Clavier is Associate Professor at Ecole Normale Supérieure de Paris, (PSL Research University), where he teaches history of metaphysics and philosophy of religion. He has written a higher PhD on the topic of creation out of nothing.

Riccardo Fedriga is Associate Professor in the Department of Philosophy and Communication Studies of the University of Bologna. Its areas of research are Medieval Philosophy and Late XIV century Theology and Metaphysics; Intentionality and Object of Thought in the History of Ideas; Social Ontology, Free Will, Censorship and Archiving Contents in the Digital Ecosystems. He published (ed. with Umberto Eco) *La filosofia e le sue storie* (Laterza 2014-2015, III voll. – 1500 p.). He's also author of several papers for National and International Journals and Publishers (Laterza, Il Mulino, Brill, Springer). His recent books are *La sesta prosa (Foreknowledge, Free Will and Contingency)*, Mimesis, Milan 2015; and *Metter le brache al mondo. Compatibilismo, conoscenza e libertà (A Study of the Compatibilism and Free Wwill in some Historical Reconstruction of the Problem)*, Jaca Book, Milan 2016.

Christian Kanzian is a Professor of Philosophy at the University of Innsbruck, Austria. He is, since 2006, the president of Austrian Ludwig Wittgenstein Society (Österreichische Ludwig Wittgenstein Gesellschaft). Kanzian's main interests are the analytic philosophy, the history of philosophy, and ontology. He is the author of four books: *Originalität und Krise – Zur systematischen Rekonstruktion der Frühschriften Kants* (1994), *Grundprobleme der Analytischen Ontologie* (with E. Runggaldier, 1998), *Ereignisse und andere Partikularien* (2001), *Ding – Substanz –*

https://doi.org/9783110594164-014

Person. Eine Alltagsontologie (2009), and an editor or co-editor of eleven books. He has also published over seventy articles.

Tomasz Kąkol is an Assistant Professor of Philosophy at the University of Gdańsk, Poland. He earned his PhD in philosophy from the University of Toruń under the direction of Prof. Urszula M. Żegleń. Kąkol's main fields of research are: ontology, philosophical logic, philosophy of mind, bioethics. Of his publications, the most noteworthy are the following: "A Formal Analysis of Selected Proofs by Aquinas for the Uniqueness of God" (in *Substance and Atiribute. Western and Islamic Traditions in Dialogue*, edited by Ch. Kanzian and M. Legenhausen Ontos Verlag 2007), "The Same P-Relation as a Response to Critics of Baker's Theory of Constitution" (in *The Journal of Philosophical Logic*, vol. 34 (2005)), "Against substantialism" (in *Filozofia Nauki* (*Philosophy of Science*), 4 (2010)), "Spinozian argument for numerical substantial monism" (in *Filozofia XVII wieku – twórcy, problerny, kontymacje* (*The Philosophy of the 17th Century - Its Authors, Issues and Continuations*), edited by J. Żelazna, Toruń: Wydawnictwo Naukowe UMK, 2011).

Zbigniew Król is a philosopher who works at the Department of the Philosophy of Science, Sociology and Foundations of Technology, Faculty of Administration and Social Sciences, Warsaw University of Technology, and at the two departments of the Institute of Philosophy and Sociology, Polish Academy of Sciences: the Department of Inquiries on Ancient Philosophy and the History of Ontology. His research concerns hermeneutics, the hermeneutical philosophy of mathematics and science, the history of mathematics and science, logic, mathematics, ontology and epistemology. He is an author of many papers and three monographs: *Plato and the Foundations of Modern Mathematics: the Concept of Number by Plato* (2005), *Mathematical Platonism and Hermeneutics* (2006) and *Platonism and the Development of Mathematics. Infinity and Geometry* (2015). He is a dean of the Department of Administration and Social Sciences of the Warsaw University of Technology.

Uwe Meixner is a professor of philosophy at the University of Augsburg, Germany. He previously was bound up with the University of Regensburg (till 2010) and earned his PhD from that University in 1986, under the direction of Franz von Kutschera. His main fields of research are Theoretical Philosophy (especially, Logic, Metaphysics, and Philosophy of Mind) and the History of Philosophy. The places of his visiting professorships or research projects are: Innsbruck, Mainz, München, Münster in Westfalen, Notre Dame (USA), Osnabrück, Regensburg, Saarbrücken, and Salzburg. Uwe Meixner is the author of eleven books: *Handlung, Zeit, Notwendigkeit: Eine ontologisch-semantische Untersuchung* (1987), *Ereignis und Substanz: Die Metaphysik von Realität und Realisation* (1997), *Axiomatic For-*

mal Ontology (1997), *Theorie der Kausalität: Ein Leitfaden zum Kausalbegriff in zwei Teilen* (2001), *The Two Sides of Being: A Reassessment of Psycho-Physical Dualism* (2004), *Einführung in die Ontologie* (2004), *David Lewis* (2006), *The Theory of Ontic Modalities* (2006), *Modalität: Möglichkeit, Notwendigkeit, Essenzialismus* (2008), *Philosophische Anfangsgründe der Quantenphysik* (2009), *Modelling Metaphysics: The Metaphysics of a Model* (2010). He has published over 100 articles, and is a co-editor of two journals: *Logical Analysis and History of Philosophy* and *Metaphysica*.

Elisa Paganini is Lecturer in Philosophy at the University of Milan (Università degli Studi di Milano). Her principal research interests are in philosophy of language and metaphysics. She is the author of two books: *La realtà del tempo* [*The Reality of Time*] (2000) and *La vaghezza* [*Vagueness*] (2008). Her published papers have addressed philosophy of time and vagueness; they have appeared in *Philosophical Studies*, *Erkenntnis*, *Dialectica*, and other journals.

Mirosław Szatkowski is a professor of philosophy at the Warsaw University of Technology, Poland. He earned his PhD in philosophy from Jagiellonian University in Cracow, Poland, and was habilitated at the Ludwig-Maximilians University in Munich, Germany. Szatkowski's main fields of research are: logic, the foundations of mathematics, and formal ontology. In these areas, he has published papers in the following professional journals: *Studia Logica*, *Zeitschrift für mathematische Logik und Grundlagen der Mathematik* (*Mathematical Logic Quarterly*), *Archiv für Mathematische Logik und Grundlagenforschung* (*Archive for Mathematical Logic*), *Notre Dame Journal of Formal Logic*, *Journal of Applied Non-Classical Logics*, *Journal of Logic, Language and Information*, and *Metaphysica*; and in several collective volumes. He has edited five volumes: *Ontological Proofs Today* (Frankfurt: Ontos Verlag, 2012), *Dualistic Ontology of the Human Person* (München: Philosophia, 2013), *Substantiality ana Causality* (Boston/Berlin/Munich: Walter de Gruyter, 2014), *God, Truth, and other Enigmas* (Boston/Berlin/Munich: Walter de Gruyter, 2015), and *Analytically Oriented Thomism* (Editiones Scholasticae, 2016).

Alfredo Tomasetta is research fellow at the School of Advanced Studies IUSS Pavia. His principal research interests are in philosophy of mind and analytic metaphysics. He published articles in *Theoria* and the *Polish Journal of Philosophy*, and authored three books: *Esistenza necessaria e oggetti possibili* [*Necessary Existence and Possible Objects*] (2008), *Coscienza e modalità* [*Consciousness and Modality*] (2012) and *Persone umane* [*Human Persons*] (2015).

Abstracts

Andrea C. Bottani and Riccardo Fedriga
"Ockham, Plantinga and the Row of Ants"

For millennia, philosophers have discussed whether divine omniscience is compatible with human freedom – conceived of in a libertarian way – or not. If libertarianism is true, some actions are free and no action is free unless it is within the agent's power to act otherwise. If God is omniscient, however, he completely foreknows how I will act in the future, which seems to entail that it is never within my power to act otherwise, provided I cannot change God's past beliefs. Therefore, I am not free in the libertarian sense.

Ockham famously contrasted this conclusion. According to him, propositions about God's past foreknowledge of future human actions are not strictly but only "by word" (secundum vocem) about the past. In Nelson Pike's more recent terms, they describe "soft" rather than "hard" facts. Soft facts about the past fail to be "accidentally necessary", so it is within our power to act in such a way that God would not have believed what in fact he does believe. In "Ockham's Way Out" Plantinga made efforts to clarify the point. Even in his version, however, Ockham's way-out faces a number of problems.

Our aim is to defend Ockham's way-out by defining in new terms both the notion of a hard fact and the idea that we are, in some sense, able to do otherwise. We propose to interpret the notion of a hard fact in terms of grounding, and identify accidentally necessary facts with a proper subset of hard facts, making of accidental necessity a non-modal property. By contrast, we characterize our power to do otherwise in genuinely modal terms. We argue that, arranged that way, Ockham's way-out is by and large more viable than in Plantinga's version.

Paul Claver
"The importance of Being Timeless"

In this paper we run through some of the arguments pro and contra God's timelessness. The temporalist view is supposed to provide us with an analogical concept of divine person, whereas divine timelessness makes God lacking an essential feature of godhead such as being a true living person interacting with his creatures in a genuine dialogue.

I suggest another assessment of the balance of the arguments and try to emphasize the threefold importance of being timeless 1°) it solves the problem of in-

https://doi.org/9783110594164-015

compatibility between divine foreknowledge and libertarian freedom better than the temporalist view, by cancelling the concept of foreknowledge. 2°) God's timeless omniscience has not the scandalous consequences in theodicy it is often associated with. 3°) God's timeless omniscience does not undermine the autonomy of creatures.

Christian Kanzian
"Temporal Relations as Epiphenomena"

The aim of my article is to sketch out a theory of time, or more specifically, of temporal relations, since I am primarily focused on McTaggert's B-series. I contend that temporal relations are epiphenomena, constituted by events. In order to make this thesis and its implications plausible it is necessary to present a framework of categories in which events have their proper place, and to make clear how events could be the constitutional basis of temporal relations. The key idea concerning constitution is that constitution is a formal tie, which implies the epiphenomenal status of the constituted.Finally I try to give an application of my theory on God and time, respectively on infinity and time.

Tomasz Kąkol
"In Defense of Presentism and Extratemporal God"

Presentism is the stance defended by the minority among contemporary ontologists of time, whereas extratemporal God is often thought of as the opposite of "the living God". In my paper I try to defend presentism using two positive and two negative arguments: 1) the argument from its presence (nomen-omen) in contemporary physics; 2) the argument from explanation; 3) the argument from the failures of the alternative strategies (a usual explanation of the intuition of time flow, the entropic theory of time, the causal theory of time); 4) the argument from the rebuttal of the alleged "bad" consequences of presentism (relative existence, "how fast does time flow" objection, rejection of the so-called principle of verifier). Although I owe those arguments to Jerzy Golosz, the ontology of time I propose is more Ingardenian in spirit and more moderate: in particular, presentism is here local, contingent and doesn't entail endurantism. As regards God, we have the mathematical model of the tensless dynamics (Michal Heller) that not only answers the objection I mentioned but also makes sense of the traditional "dark" conception of *creatio continua*.

Srećko Kovač
"Concepts, Space-and-Time, Metaphysics (Kant and the dialogue of *John* 4)"

Kant's theory of transcendental ideas can be conceived as a sort of model theory for an empirical first-order object theory. The main features of Kant's theory of transcendental ideas (especially its antinomies and their solutions) can be recognized, in a modified way, in a religious discourse as exemplified in the dialogue of Jesus and the Samaritan woman (John 4). In this way, what is by Kant meant merely as regulative ideas obtains a sort of objective reality and becomes a religiously founded metaphysics. A metaphysical theory of a religious dialogue is formalized on the basis of an extended justification logic of evidence and wish agents.

Zbigniew Król
"Basic Intuitions Concerning the Concept of Infinity in Mathematics from the Historical and Theological Point of View"

The basic strategies of defining the concept of actual infinity in mathematics are analyzed in this paper together with the relevant historical, philosophical and theological contexts. Actual infinity arises in mathematics when the so-called "God's point of view" is applied and analyzed. There are two main possibilities regarding the introduction of this concept in modern set theory. The first is connected with the upward construction of an infinite set and the set containing an inductive set. The second defines an infinite set using non-well-founded sets. Some other ways to use the concept of infinity along with actually infinite objects are also presented.

Uwe Meixner
"No Life without Time"

This paper begins by distinguishing concepts of existence, and moves on to pointing out that life and aliveness is essentially time-dependent. Since God is essentially a living God, this makes God essentially time-dependent. The paper concludes by showing how the conclusion can be avoided that God is essentially dependent on Creation (by dint of being essentially dependent on Time).

Ludwig Neidhart
"God and Time. A defense of God's timelessness"

In this paper I want to demonstrate how the traditional doctrine of Divine timelessness can be inferred from classical theistic principles. My main argument focuses on the concept of God provided by the so-called Perfect Being Theology. I also reflect upon three other lines of argumentation for God's timelessness and finally I take into consideration how to deal with the main counter-arguments.

Francesco Orilia
"The Moral Desirability of Presentism"

In a presentist world there are no past events and thus *a fortiori* no past painful events. Presentists have thus argued that relief is appropriate only from their point of view and have appealed to this claim to back up their ontology. There are well-known anti-presentists strategies to counter this move. Nevertheless, one can still argue that the rejection of past painful events that comes with presentism makes this doctrine morally superior to non-presentists world views. If so, at least for those who think that the presentist v. non-presentist dispute cannot be decided on purely theoretic grounds, or for those who endorse a theistic perspective, or at least believe in the ontological efficacy of values, there is a reason in favor of the truth of presentism.

Peter Øhstrom
"Thoughs on Time, Truth and Transcendence"

The founding father of modern temporal logic, A.N. Prior, held that there is a logical tension between the Christian doctrines of human freedom and divine foreknowledge. He argued that future contingents cannot be true now, since there is no way to settle them now. In consequence, he found that the classical doctrine of divine foreknowledge has to be rejected. In this paper, it is shown that this argument can be turned around – i.e., if we hold that there are true future contingents, then we have to accept that their truths at least in part rely on some kind of transcendence that makes it possible to assume that even future contingents can be settled. This alternative argument supports the classical views held by William of Ockham and Luis de Molina.

Elisa Paganini
"McTaggart, Lewis and the Problem of Temporary Intrinsics"

McTaggart's Paradox has been considered a special case of Lewis's Problem of Temporary Intrinsics (see Craig (1998), Rea (2003) and Rettler (2012)). I argue instead that the Problem of Temporary Intrinsics cannot simply be applied to the Problem of the passage of time and therefore that McTaggart's Paradox cannot be a special case of the Problem of Temporary Intrinsics. This observation is relevant in order to point out the difference between the change in objects or events over time (i.e. the subject of Lewis's Problem) and the change (or passage) of time (i.e. the subject of McTaggart's Paradox).

Alfredo Tomasetta
"Is Dualism Compatible with Classical Theism?"

In this paper I discuss the relation between dualism in the philosophy of mind and Classical Theism. In particular, I argue that a number of important contemporary dualists in the philosophy of mind subscribe to a thesis that is incompatible with classical theistic belief. I first state the thesis – which I label 'Thesis T'. Then I show that Thesis T is explicitly endorsed (or, at the very least, clearly suggested) by influential contemporary dualists in the philosophy of mind. Finally, I present a short argument to the effect that Thesis T and Classical Theism are indeed incompatible, and conclude by exploring the logical space of possible reactions to the argument.

Kazimierz Trzęsicki
"In What Sense is God Infinite? A Consideration in Historical Perspective"

Conceptions of the infinity of God by scholars who influenced Cantor's idea of Absolute Infinite are discussed. The common features of these conceptions are signalized. It is stressed that concepts and tools of mathematics are proper instruments of modelling and better understanding the infinity of God. We try to justify the thesis that Cantor's set theory depends on a theological conception of an infinite and rational God in an analogical way as the idea of zero depends on the idea of positive nothing.

Person Index

https://doi.org/9783110594164-016

Subject Index

https://doi.org/9783110594164-017